1

Rudolf Steiner's
Nature Stories

edited by Glen Atkinson (PhD)

Many Thanks to the Rudolf Steiner Archive
for having these materials available for us all.

Agriculture Course is the 1938 Lili Kolisko edition
as she may have edited it.

© Garuda Consultants Ltd   3.2.25 v10

| Print ISBN | 978-1-06705-552-3 |
| EBook ISBN | 978-1-06705-553-0 |

Contents

Rudolf Steiner's Nature Stories

Introduction

After 40 years of delving into Dr Steiners (RS) Agriculture and Medical lectures, working out what his story is, I got to wondering what it was he actually said about how a plant grows. So, I entered 'Plant' into the search function of the Rudolf Steiner archive , and up came a range of references, of which around 20 provided some direct story of his view of how plants grow. Some of his stories I knew but a couple I had not come across before. It was good to find the 'Sap' Story, given on 31 October 1923, which tells of specific plant growth processes through the seasons. This story is then reflected into 'The Man as Symphony' (lecture 7 - 2 November 1923) , which provides a story of Elementals and the Ethers, running through the same cycle. Together these two lectures give us something very special, to add to the Agriculture Course. There are several other lecture series on the seasons, mostly given in 1923, which are more 'Christian Anthroposophical' in tone. They are useful to reference but probably a little much for the average farmer.

All the lectures and paragraphs I found relevant are collected together, as a summary in 'The Layers of the Story' essay, while the original study text is available at http://garudabd.org/dr-steiners-plant-story/ (if you have more references please let me know) .

In the first chapter I have attempted to condense all the various bits into a cohesive story through the seasons. As is RS's way, each reference tells a slightly different part of the story, from all the others.

Our challenge is to layer these stories over each other, using the obviously common denominators, as the 'structural' pieces, and merge them together to find his overall message, for how we can use nature forces to grow healthy plants, animals and humans. Altogether this story provides a seasonal based framework, upon which our Biodynamic understandings and practices, can be organised and planned. The cycle of the natural year is the subject, but it is RS tales of what goes on along the way that is the treasure. It is fair to say there are eight main stories, and these developed over time. The ninth is my contribution.

(1) The Three Worlds

(2) Direct and Indirect growth processes

(3) The polarity between the 'Sal' processes of the Earth and the 'Sulf' processes in the above ground

(4) The growth processes from the previous year hold over and influence the following season.

(5) The 'Sap' seasonal stories of Wood Sap, Life Sap and Cambium and the Ethers and elementals.

(6) The Cosmic and Earthly, Forces and Substances

(7) The primary Energetic Bodies role in Nature's growth

(8) The Seed Chaos

(9) Alchemical Chemistry

The second chapter tells the story of the four Primary Energetic Bodies' direct influence of nature, while the third chapter explores the Physical Formative Forces role in plant growth and animal husbandry. These activities are the primary Bodies working just behind the physical substances, and are one of the main conversations in the Agriculture Course, out of which four new Preparations are suggested.

The final chapter looks into the question of when 'seed chaos' occurs.

Ultimately, each of these stories has to be woven together into a fabric we see as Plant Growth. As with cloth, the Fabric of Nature is made of threads of energy, that can be pulled and released. RS's gift has been to show us the substances that control these threads. Through this journey we are drawn into a 'Science of the Circle', which many cultures have used in various ways.

The theme running through it all can be summed up in this from "The Spirit in the Realm of Plants" given in Berlin, December 8, 1910 -

"Thus we cannot picture the earth only as a physical structure, for the physical structure is for us something like our own physical body, which can be seen with the outer eyes and touched with the hands, and which is observed by outer science. This is the earth body that present-day astronomy or geology studies. Then we have to direct our attention to what in the human being we have come to know as the etheric body or life body. The earth also has such an etheric body, and it also has an astral body. This is what awakens every spring as the thoughts and feelings of the earth, which recede when winter approaches so that the earth rests in its own ego, closed off within itself, retaining only what it needs in order, through memory, to carry over the preceding into the following, retaining in the plant's seed forces what it has conquered for itself. Just as the human being, when he falls asleep, does not lose his thoughts and sensations but finds them again the next morning, so the earth, awakening again from sleep in the spring, finds the seed forces of the plants in order to permit what has been conquered in an earlier time to emerge again from the living memory of the seed forces."

These essays are part of a series 'Dr Steiners Plant Growth Stories' at the Glenopathy website.

Energetic Plant Growth – A collage of 4 lectures.

The Seasonal Complex – An Introduction to RS Plant story

The Layers of the Story – Summary of the references

The Physical Formative Forces. – 62 page collage RS texts

The Planets in the Agriculture Course. – A suggestion for the solution.

Preparations for the Seasonal Complex – Practical steps

References – 56 pages of RS paragraphs used in the summary above.

This series is in celebration of the 100 year anniversary of Rudolf Steiner's Agriculture Course – 7th June 2024, where Biodynamic Agriculture begins. This is truly a gift to humanity from one who sees things clearly.

My favorite saying of his is **"I Stood Up and Walked"**. So may we all.

Energetic Plant Growth

My edits, in italics, are to fill in the gaps and to harmonise the five or so different languages he uses throughout the different lectures. My aim is to make one sensible sequential story, throughout the Seasons.

My dear friends,

In these first lectures you must observe how all agricultural products arise; how Agriculture lives in the totality of the Universe. (3)

Let us draw the plant in its entirety (Diagram 16). Down here you have the root; up there the unfolding leaves and blossoms. And as above, in the leaves and blossoms, the astral element (red) is acquired from contact with the air, so the ego-potentiality (orange) develops below in the root through contact with the manure. The farm is truly an organism. The astral element is developed above, and the presence of orchard and forest assists in collecting it. If animals feed in the right way on the things that grow above the earth, then they will develop the right

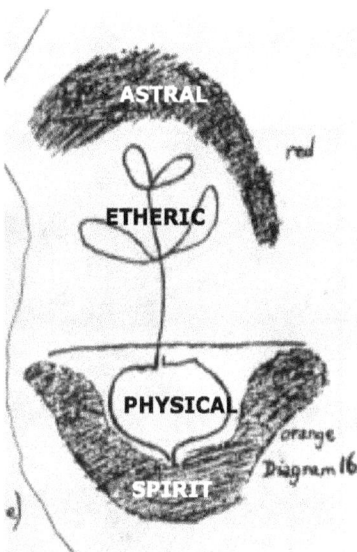

internal Spirit / ego-potentiality in the manure. If they produce, this ego-potentiality, *it will connect with the plants ego potential and Cosmic Forces, already within the Earth (see gold balls pg 11) and* work on the plant from the root, causing it to grow upwards from the root in the right way according to the forces of gravity. It is a wonderful interplay. (5)

The plant-world develops in such a way that it represents only physical corporeality, etheric corporeality; that is, in the actual plants themselves. But when we come to the astral element of the plant-world, we must imagine this astral element of the plant-world as an astral atmosphere which encompasses the earth. The plants themselves have no astral bodies, but the earth is enveloped in an astral atmosphere, and this astrality plays an important part, for instance, in the process of the unfolding of blossom and fruit. The terrestrial plant-world as a whole, therefore, has one uniform, common astral body which nowhere interpenetrates the plant itself, except at most in a very slight degree when *pollination* begins in the blossom. Generally speaking, it floats cloud-like over the vegetation and stimulates blossom and fruit formation. (8)

Plants Astrality

Let us be clear in our minds about the basis of plant-life. Let us picture the surface of the Earth and the plants growing out of it. We know that the physical organization of the plant is permeated by its ether body, *however* the plant would not be able to unfold if the all pervading astrality did not contact it from above by way of the blossom (lilac).

The plant has no astral body *within* it but the astrality touches it from above. As a rule the plant does not absorb the astrality but only allows itself to be touched by it. The plant does not assimilate the astrality but towards the blossom and the fruit there is interplay with the astrality which does not, as a rule, combine with the ether-body or physical body of the plant.

In a poisonous plant, however, it is different. In a poisonous plant the astrality penetrates into the actual substance of the plant and combines with it. A plant such as belladonna or, let us say, henbane, hyoscyamus, sucks in the astrality either strongly or more moderately and so bears astrality within itself — in an uncoordinated state, of course, for if it were coordinated the plant would have to become an animal. It does not become an animal; the astrality within it is in a compressed state. (9)

Plants Spirit / Ego

The plant egos / *Spirit* dwell in the very centre of the earth, whereas the animal group souls , *and plants astrality,* circle round the earth like trade winds. All these plant egos at the centre point of the earth are mutually interpenetrating beings, for in the spiritual world a law of penetrability prevails and all beings pass through one another. We see the animal group souls moving over the earth like trade winds, and how in their wisdom they carry out what appears to be done by the animals. Studying the plant we see that its head — the root — is directed towards the center of the earth where its group ego is to be found. The earth itself is the outward expression of soul and spirit beings. From the spiritual point of view the plants

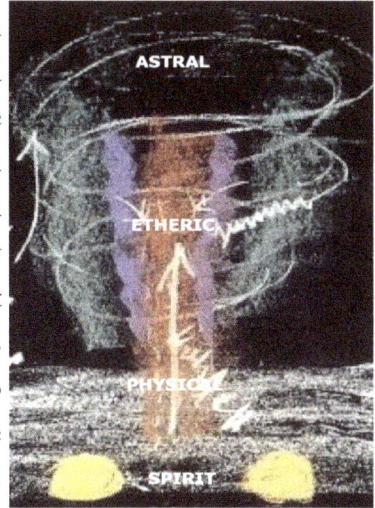

seem like the nails of our fingers. The plants belong to the earth, and when we look at them singly we do not see a complete entity, for the single plant is just one among the whole number of beings constituting a group ego. In this way we can enter into what the plants themselves feel. The part of the plant that springs up out of the earth, what from within the earth strives up to the surface, is of a different nature from what is growing under the earth. There is a difference between the cutting off of blossoms, stalk, leaves, and the tearing up of a root. The former gives the plant soul a feeling of well-being, of pleasure, just as it gives pleasure to a cow, for example, when the calf sucks milk from her udder. There is actual similarity between the milk of animals, and that part of a plant which pushes its way out of the earth. When in late summer we go through fields where corn is being cut, where the blade is passing through the corn stems, then the whole fields breathe out a feeling of bliss. It is an intensely significant moment when we not only watch the reaping with our physical eyes, but perceive the feeling of contentment sweeping over the earth as the corn falls to the ground. But when the roots of the plants are pulled up, then that is painful for the plant souls. (6)

Curiously enough, the spiritual investigator becomes aware that it is generally impossible to consider the world of plants, this wonderful covering of the earth, as something existing by itself. When confronted with the plant he feels just as he does regarding a finger, which he can consider only as belonging to a complete human organism. The plant world cannot be considered in isolation, because to the view of the spiritual investigator the plant world at once relates itself to the entire planet earth and forms a whole with the earth, just as the finger or piece of bone or the brain forms a whole with our organism. And whoever merely looks at plants by themselves, remaining with the particular, does the same as one who wishes to explain a hand or a piece of human bone by itself. The common nature of plants simply cannot be considered in any other way than as a member of our common planet earth.

An outer circumstance might already suggest to us that, just as every stone has a certain relationship to the earth, so also everything plant-like belongs to it. Just as every stone, every lifeless body, shows its relationship to the earth by being able to fall onto the earth, where it finds a resistance, so every plant shows its relationship to the earth by the direction of its stem, which is always such that it passes through the center of the earth. All stems of plants would cross at the earth's center if we extended them to that point. This means that the earth is able to draw out of its center all those *Cosmic* force radiations that allow the plants to arise.

If we now study grain-producing plants, we discover remarkable little organs present in all these plants. Small structures in the starch cells are discovered. These cells are constructed in quite a remarkable way, so that within them there is something like a loose kernel. These structures have the unique property that the cell wall remains insensitive to the kernel at only one spot. If the kernel slips to another spot, it touches the cell wall, leading the plant to return to its earlier position. Such starch cells are found in all plants whose main orientation is toward the center of the earth, so that the plant has an organ within that always makes it possible for it to direct itself in its main orientation toward the center of the earth. This discovery, made during the nineteenth century by various scientists, is certainly wonderful, and it is most remarkable if it is simply presented as it is. Natural science *also* shows us that mistletoe does not have those curious starch cells that orient the plant toward the center of the earth.

But now let us turn to something else. If the leaf of a plant is studied, it is discovered that the outer surface is actually always a composite of many small, lens-like structures, similar to the lens in our eye. These 'lenses' are arranged in such a way that the light is effective only if it falls onto the surface of the leaf from a very specific direction. If it falls from another direction, the leaf instinctively begins to turn in such a way that the light can fall into the center of the lens, because when it falls to the side it works in another way. Thus there are organs for light on the surface of the leaves of plants. These light organs, which actually can be compared with a kind of eye, are spread out over the plants, but the plant does not see by means of them; rather the sun being looks through them to the earth being. These light organs bring it about that the leaves of the plant always have the tendency to place themselves perpendicularly to the sunlight.

In this — in the way the plant surrenders itself to the sun's activity in spring and summertime — we have the plant's second main orientation. The first orientation is that of the stem, through which the plants reveal themselves as belonging to the earth's self-consciousness; the second orientation is the one through which the plants express the earth's surrender to the activity of the sun beings.

You will find little by little how the plant covering of our earth is the sense organ through which earth spirit and sun spirit behold each other. The carbohydrates can arise only if the sun spirit and the earth spirit kiss through the plant being.

Now, if we are able to consider this earth organism from a spiritual scientific viewpoint, we can go still further. We can ask ourselves, what is the situation with the earth organism as such?

Summer and Winter

In studying an organism we know that alternations of different conditions are revealed. The human and animal organisms reveal a waking and a sleeping condition alternating in time. Can we, from a spiritual scientific viewpoint, find something similar regarding the body of the earth, the earth organism? To outer consideration, what follows may appear to be a

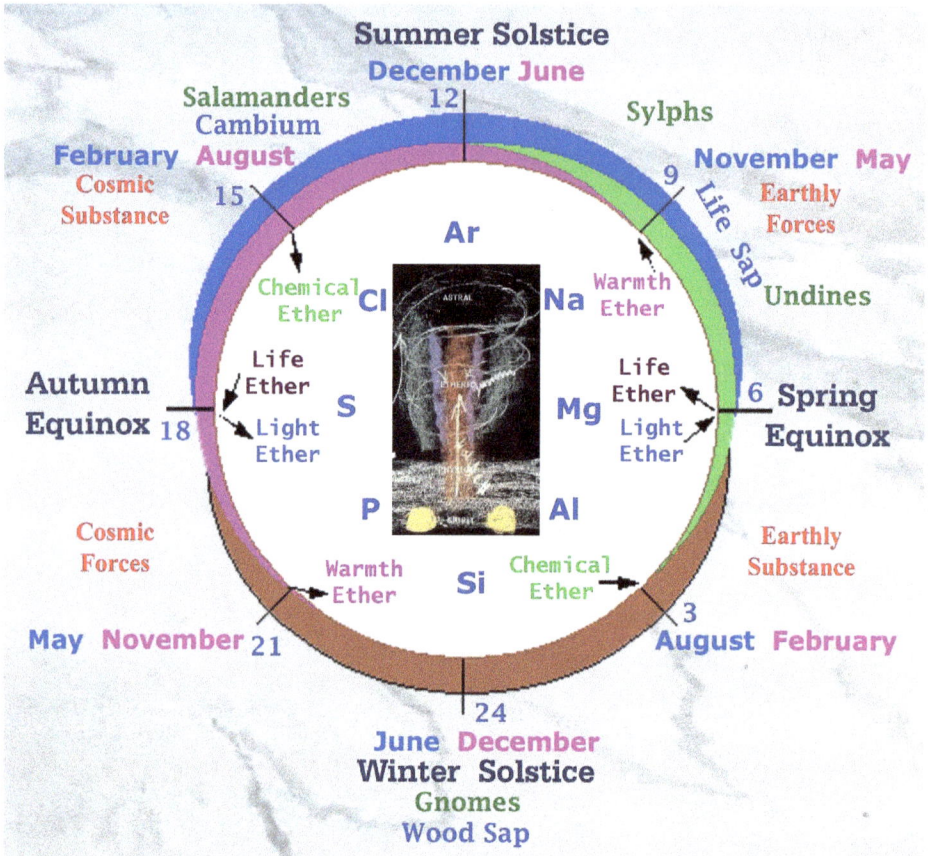

Summer Solstice
December June
12

Salamanders
Cambium

February August
Cosmic Substance
15

Sylphs

November May
Earthly Forces

9
Life Sap

Ar

Chemical Ether — Cl

Na — Warmth Ether

Undines

Life Ether

S

Mg

Life Ether

Autumn Equinox 18

Light Ether

Light Ether

6 Spring Equinox

P

Al

Cosmic Forces

Earthly Substance

May November 21

Warmth Ether

Si

Chemical Ether

August February

3

24

June December
Winter Solstice
Gnomes
Wood Sap

mere comparison, but for spiritual research it is not a comparison but a fact. If we study the curious lawfulness of summer and winter, how it is summer on one half of the earth and winter on the other half, how this relationship alternates, and if we pay attention to how this lawfulness — as wintertime and summertime — is to be discerned in relation to all earthly life, then it will no longer appear absurd if spiritual science tells us that winter and summer in the earth organism correspond to waking and sleeping in the organisms around us. It is simply that the earth does not sleep in time in the same way as other organisms but is always awake somewhere and always asleep at some other portion of its being. Waking and sleeping move around spatially: the earth sleeps in the part where there is summer, and it is awake in the part of its being where there is winter. Thus the whole earth organism confronts us spiritually with conditions like waking and sleeping in other organisms.

The **summer** condition of the earth organism consists of a very specific relationship of the earth to the sun, and because we are dealing with a living, spirit-filled organism we may say that **it surrenders itself** to an activity that proceeds spiritually from the sun. In the **winter** condition the earth organism **closes itself off** from this sun activity, drawing itself together into itself. Now let us compare this condition with human sleep. I will now speak of what appears to be a mere analogy; spiritual science, however, provides the evidence for these observations.

If we study the human being in the evening, when he is tired, as his consciousness is diminishing, we find that all thoughts and feelings that enter our soul during the day from the outside, all pleasure and suffering, joy and pain, sink into an indefinite darkness. During this time, the human spirit being — as we have shown in the lecture about the nature of sleep[3] — passes out of the human physical body and enters the spiritual world, surrendering itself to the spiritual world. In this sleep condition it is a curious fact that the human being becomes unconscious. For the spiritual investigator (we will see how he comes to know this) it is revealed that the inner aspect of the human being, the astral body and ego, actually draw themselves out of the physical and etheric bodies, but they do not simply draw themselves out and float over him like a cloud formation; rather this whole inner aspect of the human being spreads itself out, pours itself out over the whole planetary world around us. As incredible as it may seem, it is nevertheless revealed that the human soul pours itself out in a unified way over the astral realm. The investigators who were acquainted with this realm knew well why they called what departs from the physical the 'astral body.' The reason was that this inner element draws out of heavenly space, with which it forms a unity, the forces it needs in order to replace what the day's efforts and work used up from the physical body. Thus the human being in sleep passes into the great world and in the morning draws himself back within the limits of his skin, into the small human world, into the microcosm. There, because his body offers him resistance, he again feels his ego, his self-consciousness.

This breathing out and breathing in of the soul is a wonderful alternation in human life. Sleep for Novalis means the digestion of the soul by the body. Novalis is always conscious that in sleep the soul becomes one with the universe and is digested, so that the human being can be further helped in the physical world.

With respect to his inner being, then, the human being alternates in such a way that in the daytime he draws himself together into the small world, into the limits of his skin, and then expands into the great world during the night, drawing forth through surrender forces from that world in which he is then imbedded. We will not understand the human being unless we understand him as formed out of the entire macrocosm.

For that part of the earth where it is summer, there is something similar to what goes on in the human being in the condition of sleep. The earth gives itself to everything that comes down from the sun and forms itself as it should form itself under the influence of the sun activity. In that part of the earth where it is winter, it closes itself off from the influence of the sun, lives within itself. There it is the same as when the human being has drawn together into the small, inner world, living in himself, while for the part of the earth where it is summer it is the same as when the human being is surrendered to the whole outer world.

There is a law in the spiritual world: if we direct our attention to spiritual entities far removed from one another — such as, for example, the human being here on one side and the earth organism on the other — the states of consciousness must be pictured as reversed in a certain sense. With the human being, stepping out into the great world is the sleep condition. For the earth, the summer (which one would be inclined to consider a waking condition) is something that can only be compared with the human being falling asleep. The human being steps out into the great world when he falls asleep; in summer the earth with all its forces enters the realm of sun activity, only we must be able to think of the earth and the sun as spirit-filled organisms.

In wintertime, when the earth rests within itself, we must be able to think of its condition as corresponding to the waking condition of the human being, although it may be tempting to consider winter as the earth's sleep. When we consider entities as different from one another as the human being and the earth, however, the states of consciousness appear reversed

16

in a certain way. Now, what does the earth accomplish when it is under the influence of surrender to the sun being, to the sun spirit? To have an easier comparison, we would do well to turn the concepts around now. The earth's surrender to the sun being is simply something that may be compared spiritually with the condition of the human being when he awakens in the morning and emerges out of the dark womb of existence, out of the night, into his joys and sorrows. When the earth enters the realm of sun activity — although this could be compared with the sleep condition of the human being — all the forces that sprout forth from the earth allow the resting winter condition of the earth to pass over into the active, the living, summer condition.

What, then, are the plants in this whole web of existence? We could say that when spring approaches, the earth organism begins to think and to feel, because the sun with its being lures out the thoughts and feelings. The plants are nothing but a kind of sense organ for the earth organism, awakening anew every spring, so that the earth organism with its thinking and feeling can be in the realm of the sun activity. Just as in the human organism light creates the eye for itself in order to be able to manifest through the eye as 'light,' so every spring the sun organism creates for itself the plant covering in order to look at itself, to feel, to sense, to think by means of this plant covering. The plants cannot directly be considered the thoughts of the earth, but they are the organs through which the awakening organization of the earth in spring, together with the sun, develops its thoughts and feelings. Just as we can see our nerves emanating from the brain, developing our feeling and conceptual life through the eyes and ears together with the nerves, so the spiritual investigator sees in what transpires between earth and sun with the help of the plants the marvelous weaving of a cosmic world of thoughts, feelings, and sensations. The spiritual investigator finds that the earth is surrounded not merely by the mineral air of the earth, by the purely physical earth atmosphere, but by an aura of thoughts and feelings. For spiritual research the earth is a spiritual being whose thoughts and feelings awaken every spring, and throughout the summer they pass through the soul of our entire earth.

The plant world, however, which is a part of our entire earth organism, provides the organs through which our earth can think and feel. Woven into the spirit of the earth are the plants, just as our eyes and ears are woven into the activities of our spirit.

In spring a living, spirit-filled organism awakens, and in the plants we can see something that is pushed out of the countenance of our earth in some realm where it wants to begin to feel and think. Just as everything in the human being tends toward a self-conscious ego, so it is also in the realm of plants. The whole plant world belongs to the earth. I have already said that a person would be close to insanity if he did not think of how all feelings, sensations, and mental images are directed toward our ego. Similarly, everything the plants mediate during summertime is directed toward the earth's center, which is the earth ego. This should not be said merely symbolically! As the human being has his ego, so the earth has its self-conscious ego. That is why all plants strive toward the earth's center. That is why we may not consider plants by themselves but rather must consider them in interaction with the self-conscious ego of the earth. What unfolds itself as thoughts and sensations of the earth is similar to the thoughts and sensations that live in us, similar to whatever arises and disappears in us during our waking state, what lives in us astrally, if we speak from the viewpoint of spiritual science.

Thus we cannot picture the earth only as a physical structure, for the physical structure is for us something like our own physical body, which can be seen with the outer eyes and touched with the hands, and which is observed by outer science. This is the earth body that present-day astronomy or geology studies. Then we have to direct our attention to what in the human being we have come to know as the etheric body or life body. The earth also has such an etheric body, and it also has an astral body. This is what awakens every spring as the thoughts and feelings of the earth, which recede when winter approaches so that the earth rests in its own ego, closed off within itself, retaining only what it needs in order, through memory, to carry over the preceding into the following, retaining in the plant's seed forces what it has conquered for itself. Just as the human being, when he falls asleep, does not lose his thoughts and sensations but finds them again the next morning, so the earth, awakening again from sleep in the spring, finds the seed forces of the plants in order to permit what has been conquered in an earlier time to emerge again from the living memory of the seed forces.

When regarded in this way, the plants can be compared with our eyes and ears. What our senses are for us, the plants are for the earth organism. But what perceives, what achieves consciousness, is the spiritual world streaming down from the sun to the earth. This spiritual world would not be able to achieve consciousness if it did not have its sense organs in the plants, mediating a self-consciousness just as our eyes and ears and nerves mediate our self-consciousness. This makes us aware that we speak correctly only if we say that those beings who stream from the sun down to the earth, unfolding their spiritual activity, encounter from spring through summertime the being that belongs to the earth itself. In this exchange the organs are formed through which the earth perceives those beings, for the plants do not perceive. It is a superstition, shared also by natural science, when it is said that the plant perceives. The spiritual entities that belong to the earth activity and the sun activity perceive through the plant organs, and these entities direct toward the center of the earth all organs they need in order to unite them with the center of the earth. Thus what we have to see behind the plant covering are the spiritual entities that weave around the earth and have their organs in the plants. (7)

The Farm

A farm is true to its essential nature, in the best sense of the word, if it is conceived as a kind of individual entity in itself — a self-contained individuality. Every farm should approximate to this condition. This ideal cannot be absolutely attained, but it should be observed as far as possible. Whatever you need for agricultural production, you should try to posses it within the farm itself (including in the "farm," needless to say, the due amount of cattle). Properly speaking, any manures or the like which you bring into the farm from outside should be regarded rather as a remedy

19

for a sick farm. That is the ideal. A thoroughly healthy farm should be able to produce within itself all that it needs.

We shall see presently why this is the natural thing. So long as one does not regard things in their true essence but only in their outer material aspect, the question may justifiably arise: Is it not a matter of indifference whether we get our cow-dung from the neighbourhood or from our own farm? But it is not so. Although these things may not be able to be strictly carried out, nevertheless, if we wish to do things in a proper and natural way, we need to have this ideal concept of the necessary self-containedness of any farm.

You will recognise the justice of this statement if you consider the Earth on the one hand, from which our farm springs forth, and on the other hand, that which works down into our Earth from the Universe beyond. Nowadays, people are wont to speak very abstractly of the influences which work on to the Earth from the surrounding Universe. They are

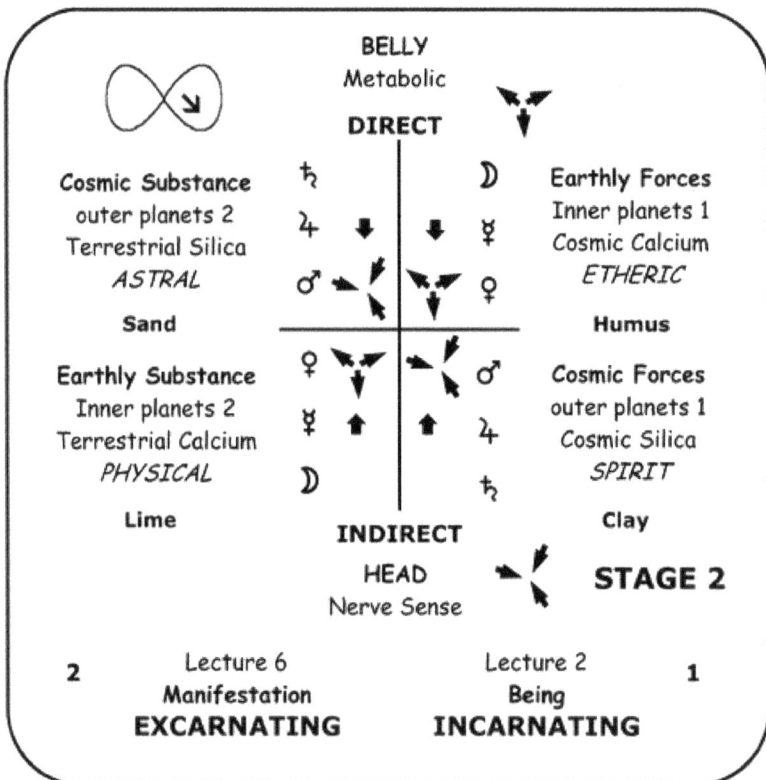

BELLY
Metabolic
DIRECT

Cosmic Substance
outer planets 2
Terrestrial Silica
ASTRAL

Earthly Forces
Inner planets 1
Cosmic Calcium
ETHERIC

Sand   Humus

Earthly Substance
Inner planets 2
Terrestrial Calcium
PHYSICAL

Cosmic Forces
outer planets 1
Cosmic Silica
SPIRIT

Lime   Clay

INDIRECT
HEAD   STAGE 2
Nerve Sense

2   Lecture 6   Lecture 2   1
     Manifestation   Being
     EXCARNATING   INCARNATING

aware, no doubt, that the Sun's light and warmth, and all the meteorological processes connected with it, are in a way related to the form and development of the vegetation that covers the soil. But present-day ideas can give no real information as to the exact relationships, because they do not penetrate to the realities involved. We shall have to consider the matter from various standpoints. Let us to-day choose this one: let us consider, to begin with, the soil of the Earth which is the foundation of all Agriculture.

I will indicate the surface of the Earth diagramatically by this line. The surface of the Earth is generally regarded as mere mineral matter — including some organic elements, at most, inasmuch as there is formation of humus, or manure is added. In reality, however, the earthly soil as such not only contains a certain life — a vegetative nature of its own — but an effective astral principle as well; a fact which is not only not taken into account to-day but is not even admitted nowadays.

But we can go still further. We must observe that this inner life of the earthly soil (I am speaking of fine and intimate effects) is different in summer and in winter. Here we are coming to a realm of knowledge, immensely significant for practical life, which is not even thought of in our time.

Taking our start from a study of the earthly soil, we must indeed observe that the surface of the Earth is a kind of organ in that organism which reveals itself throughout the growth of Nature. The Earth's surface is a real organ, which — if you will — you may compare to the human diaphragm. (Though it is not quite exact, it will suffice us for purposes of illustration). We gain a right idea of these facts if we say to ourselves: Above the human diaphragm there are certain organs — notably the head and the processes of breathing and circulation which work up into the head. Beneath it there are other organs.

If from this point of view we now compare the Earth's surface with the human diaphragm, then we must say: In the individuality with which we are here concerned, the head is beneath the surface of the Earth, while we, with all the animals, are living in the creature's belly! Whatever is above the Earth, belongs in truth to the intestines of the "agricultural individuality," if we may coin the phrase. We, in our farm, are going about in the belly of the farm, and the plants themselves grow upward in the belly of the farm. Indeed, we have to do with an individuality standing on its head. We only regard it rightly if we imagine it, compared to man, as standing on its head. With respect to the animal, as we shall presently see, it is a little different.

Why do I say that the agricultural individuality is standing on its head? For the following reason. Take everything there is in the immediate neighbourhood of the Earth by way of air and water vapours and even warmth. Consider, once more, all that element in the neighbourhood of the Earth in which we ourselves are living and breathing and from which the plants, along with us, receive their outer warmth and air, and even water. All this actually corresponds to that which would represent, in man, the abdominal organs. On the other hand, that which takes place in the interior of the Earth beneath the Earth's surface — works upon plant-growth in the same way in which our head works upon the rest of our organism, notably in childhood, but also throughout our life. There is a constant and living mutual interplay of the above-the-Earth and the below-the-Earth.

Physical Formative Forces

And now, to localise these influences, *into the Physical Formative Forces,* I beg you to observe the following. The *force* activities above the Earth are immediately dependent on Moon, Mercury and Venus supplementing and modifying the influences of the Sun. The so-called "planets near the Earth" extend their *force* influences to all that is above the Earth's surface. On the other hand, the distant planets — those that revolve outside the circuit of the Sun — work upon *the force activities* that *are* beneath the Earth's surface, assisting those influences which the Sun exercises from below the Earth. Thus, so far as *the force aspects of* plant-growth is concerned, we must look for the influences of the distant Heavens beneath, and of the Earth's immediate cosmic environment above the Earth's surface.

Once more: all that works inward from the far spaces of the Cosmos, *as Cosmic Forces,* to influence the growth of plants, works not directly — not by direct radiation — but in this way: It is first received by the Earth *from the Cosmic Substance,* and the Earth then rays it upward again. Thus, *the Cosmic Force* influences that rise upward from the earthly soil — beneficial or harmful for the growth of plants — are in reality cosmic influences rayed back again. *The Cosmic Substance is* working directly in the air and water over the Earth. The direct radiation from the Cosmos, *coming via the Cosmic Substance* is stored up beneath the Earth's surface *as Cosmic Forces* and works back from there. Now these relationships determine how the earthly soil, according to its constitution, works upon the growth of plants. (We shall take plant-growth to begin with, and afterwards extend it to the animals). (3)

With the Substance stream we must learn to distinguish those *Earthly* Forces which arise in the cosmos but are absorbed by the earth and work upon plant-growth as *Earthly Substance,* from within the earth. These forces come from Mercury, Venus and Moon and act not directly, but through the mediation of the earth. They must be taken into account if we wish to follow up how the mother-plant gives rise to a daughter- plant, and so on. On the other hand, we have to consider the *Cosmic Substance* forces taken by the plant from the outer-earthly, and brought to it by way of the atmosphere from the outer planets. Broadly speaking, we may say that the forces coming from the nearer planets are very much influenced by the workings of lime in the soil, while those coming from the distant planets fall under the influence of silicon. And, in fact, *the Cosmic Force* workings of silicon, even though they proceed from

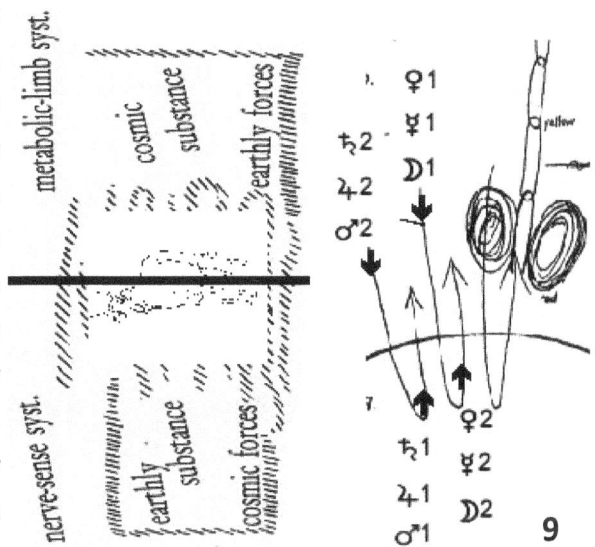

23

the earth, act as mediators of the forces coming from Jupiter, Mars and Saturn, but not for those of Moon, Mercury and Venus, *which also work in the Earth, as Earthly Substance.*

Now I want you to imagine that Diagram No. 9 represents the earth level, where the influences of Venus, Mercury and Moon; enter *as Earthly Forces* into the earth and stream again from below upwards as Earthly Substance. These are the forces which cause the plant to grow during the season, later produce the seed, and by means of this seed a new plant', a second plant, then yet a third and so on. (I indicate this schematically). All this goes into the power of reproduction and streams on into the succeeding generations. The *Cosmic Substance* forces, however, which take the other path, remaining above the earth level, come from the *secondary* distant planets. I can draw this schematically in this way. These forces cause the plant either to spread into its surroundings or to become fat and juicy, to build matter into itself such as we can use for food because it is produced again and again in a continuous stream.

Take for example the flesh of fruit - an apple or a plum - which we can break off and eat; all this is due to the *secondary* workings from the distant planets. (4)

Consider the earthly soil. To begin with, we have those *Cosmic Substance* influences that depend on the farthest distances of the Cosmos — the farthest that come into account for earthly processes. These effects are found in what is commonly called sand and rock and stone. Sand and rock — substances impermeable to water, which, in the common phrase, "contain no foodstuffs" — are in reality no less important than any other factors. They are most important for the unfolding of the growth-processes, and they depend throughout on the influences of the most distant cosmic *substance* forces. And above all — improbable as it appears at first sight — it is through the sand, with its silicious content, that there comes into the Earth what we may call the life-ethereal and the chemically influential elements of the soil. These influences then take effect as they ray upward again from the Earth, *as Cosmic Forces.*

The way the soil itself grows inwardly alive and develops its own chemical processes, depends above all on the composition of the sandy portion of the soil. What the plant-roots experience in the soil depends in no small measure on the extent to which the cosmic life and cosmic chemistry are seized and held by means of the stones and the rock, which may well be at a considerable depth beneath the surface. Therefore, wherever we are studying plant growth, we should be clear in the first place as to the geological foundation out of which it arises. For those plants in which the root-nature as such is important, we should never forget that a silicious ground — even if it be only present in the depths below — is indispensable. I would say, thanks be to God that silica is very widespread on the Earth — in the form of silicic acid, for instance, and in other compounds. It constitutes 47-48% of the surface of the Earth, and for the

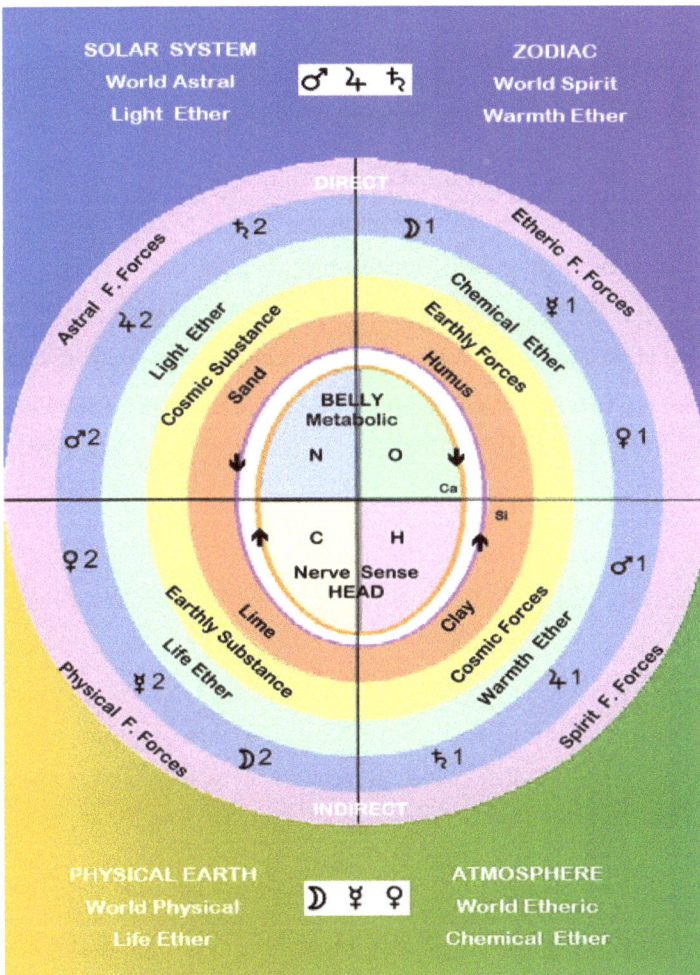

SOLAR SYSTEM — World Astral — Light Ether — ♂ ♃ ♄ — ZODIAC — World Spirit — Warmth Ether

DIRECT

♄2 — ☽1 — Etheric F. Forces — Astral F. Forces — ♃2 — Chemical Ether — ☿1 — Light Ether — Cosmic Substance — Earthly Forces — Sand — Humus — BELLY Metabolic — N — O — Ca — ♂2 — ♀1 — Si — C — H — Nerve Sense HEAD — ♀2 — Clay — Earthly Substance — Lime — Cosmic Forces — Warmth Ether — Life Ether — ♂1 — Physical F. Forces — ☿2 — ♃1 — Spirit F. Forces — ☽2 — ♄1

INDIRECT

PHYSICAL EARTH — World Physical — Life Ether — ☽ ☿ ♀ — ATMOSPHERE — World Etheric — Chemical Ether

25

quantities we need we can reckon practically everywhere on the presence of the silicic activity.

But that is not all. All that is thus connected, by way of silicon, with the root-nature, must also be able to be led upward through the plant. It must flow upward. There must be constant interaction between what is drawn in from the Cosmos by the silicon, *as Cosmic Forces* and what takes place — forgive me! —in the "belly" up above; as *Cosmic Substance* for by the latter process the "head" beneath must be supplied with what it needs. The "head" is supplied *with Cosmic Forces* out of the Cosmos, but it must also be in mutual interaction with what is going on *with the Cosmic Substance* in the "belly," above the Earth's surface. In a word, that which pours down from the Cosmos and is caught up beneath the surface, *as Cosmic Forces,* must be able to pour upward again. And for this purpose is the clayey substance in the soil. Everything in the nature of clay is in reality a means of transport, for the influences of cosmic *force* entities within the soil, to carry them upward again from below.

When we pass on to practical matters, this knowledge will give us the necessary indications as to how we must deal with a clayey soil, or with a silicious soil, according as we have to plant it with one form of vegetation or another. First we must know what is really happening. However else clay may be described, however, else we may have to treat it so as to make it fertile — all that, no doubt, is most important in the second place, but the first thing is to know that clay is the carrier of the cosmic upward stream.

But this up-streaming of the cosmic influences is not all. There is also the other process which I may call the terrestrial or earthly — that process which is *also* going on in the "belly" and which depends on a kind of external "digestion." For plant-growth, in effect, all that goes on through summer and winter in the air above the Earth is essentially a kind of digestion. All that is thus taking place through a kind of digestive process, must in its turn be drawn downward into the soil. Thus a true mutual interaction will arise with all the forces and fine homeopathic substances which are engendered by the water and air above the Earth. All this is drawn down into the soil by the greater or lesser limestone content of the soil. The limestone content of the soil itself, and the distribution of limestone substances in homeopathic dilution immediately above the soil — all this is there to carry into the soil the immediate *Earthly Forces* process.

In due time there will be a science of these things — not the mere scientific jargon of to-day — and it will then be possible to give exact indications. It will be known, for instance, that there is a very great difference between the warmth that is above the Earth's surface that is to say, the warmth that is in the *Earthly Force* domain of Sun, Venus, Mercury and Moon — and that warmth which makes itself felt within the Earth; which is under the *Cosmic Force* influence of Mars, Jupiter and Saturn. For the plant, we may describe the one kind as leaf-and-flower warmth, and the other as root warmth. These two warmths are essentially different, and in this sense, we may well call the warmth above the Earth dead, and that beneath the Earth's surface living.

The warmth beneath the Earth decidedly contains some inner principle of life, *that it gains from contact with the Earth's Etheric body*. It is alive; moreover in winter it is most of all alive. If we human beings had to experience the *etherized* warmth which works within the Earth, we should all grow dreadfully stupid, for to be clever we need to have dead warmth brought to our body. But the moment the warmth is drawn into the Earth by the limestone-content of the soil, or by other substantialities within the Earth — the moment any outer warmth passes over into inner warmth — it is changed into a certain condition of vitality, however delicate.

People to-day are well aware that there is a difference between the air above the soil and the air within, but they do not observe that there is also this difference between the warmth above and within. They know that the air beneath the surface contains more carbonic acid, and the air above, more oxygen, but again they do not know the reason. The reason is that the air too is permeated by a delicate vitality the moment it is absorbed and drawn into the Earth's *Etheric activity*.

So it is both with the warmth and with the air; they take on a slightly living quality when they are received into the Earth. The opposite is true of the water and of the solid earthy element itself. They become still more dead inside the Earth than they are outside it. They lose something of their external life. Yet in this very process they become open to receive the most distant cosmic forces. (3)

To the outwardly perceptible, visible world *of Substance* there belongs the invisible world *of Force* and these, taken together, form a whole. The marked degree to which this is the case first appears in its full clarity when we turn our attention away from the animals to the plants.

Plant-life, as it sprouts and springs forth from the earth, immediately arouses our delight, but it also provides access to something which we must feel as full of mystery. In the case of the animal, though certainly its will and whole inner activity have something of the mysterious, we nevertheless recognize that this will is actually there, and is the cause of the animal's form and outer characteristics. But in the case of the plants, which appear on the face of the earth in such magnificent variety of form, which develop in such a mysterious way out of the seed with the help of the earth and the encircling air — in the case of the plant we feel that some other factor must be present in order that this plant-world may arise in the form it does.

The earthly and cosmic forces work in the processes of agriculture through the substances of the Earth. You know that in terms of contemporary chemistry, the main ingredients of albumen are the four main natural substances, carbon, oxygen nitrogen and hydrogen, and , in addition, sulphur, as, so to speak, a omnipresent mediator, and homeopathic agent in the operations of the other four. *In the great spheres of nature we can identify, Hydrogen as the dominant chemical element of the cosmic spaces, populated by the stars. Nitrogen is found concentrated in the atmospheres of some planets, with our own atmosphere comprising 80% nitrogen. Oxygen, we find only in our atmosphere, at 20%, as a expression of the very life forms it helps to support, while our Earthly forms are primarily Carbon based.*

What interests us here is the fact that the function performed in the

28

external world by C,H,O,N and their mediator sulphur is , *the same activity as* is being individualized in man through the four organic systems. You will see then that the *Spirit inspired* Ego organisation is connected with the Hydrogen in the same way that the Physical organisation is connected with Carbon, the Etheric organisation with Oxygen and the Astral organisation with Nitrogen. The composition of the external atmosphere is of such a nature as to furnish the ratio for the connection between the astral and etheric bodies and concurrently between their partners the physical body and ego. *Where these chemicals are found, we see the action of the energetic activities* (10)

Elementals

When spiritual vision is directed to the plant-world, we are immediately led to a whole host of beings, which were known and recognized in the old times of instinctive clairvoyance, but which were afterwards forgotten and today remain only as names used by the poet, names to which modern man ascribes no reality. To the same degree, however, in which we deny reality to the beings which whirl and weave around the plants, to that degree do we lose the understanding of the plant-world. This understanding of the plant-world, which, for instance, would be so necessary for the art of healing, has been entirely lost to present-day humanity.

We have already recognized a very significant connection between the world of the plants and the world of the butterflies; but this too will only come rightly before our souls when we look yet more deeply into the whole weaving and working of plant-life.

Plants send down their roots into the ground. Anyone who can observe what they really send down and can perceive the roots with spiritual vision (for this he must have) sees how the root-nature is everywhere surrounded, woven around, by elemental nature spirits. And these elemental spirits, with an old clairvoyant perception designated as gnomes and which we may call the root-spirits, can actually be studied by an imaginative and inspirational world-conception, just as human life and animal life can be studied in the sphere of the physical. We can look into

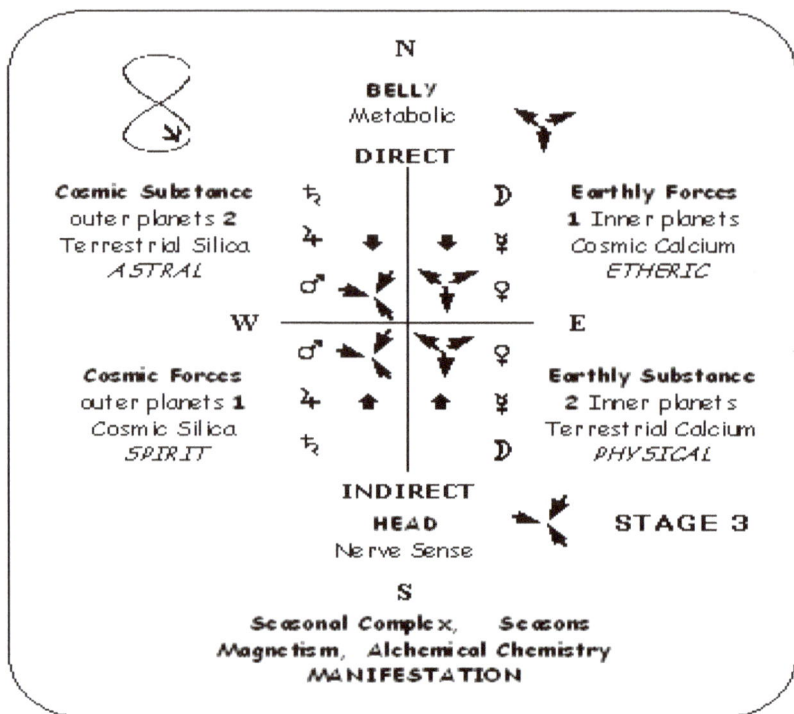

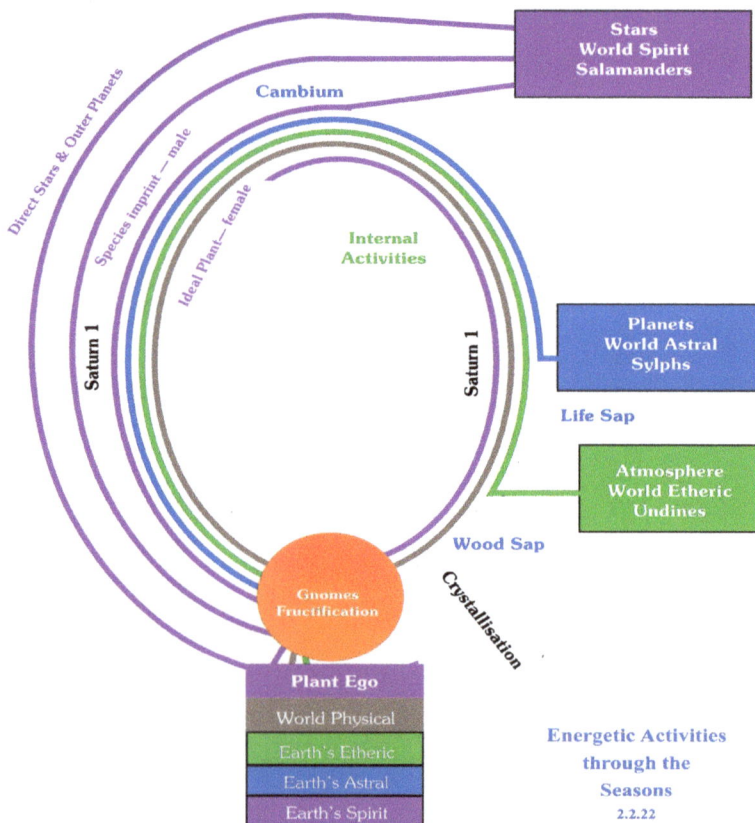

the soul-nature of these elemental spirits, into this world of the spirits of the roots.

These root-spirits, are, so to say, a quite special earth-folk, invisible at first to outer view, but in their effects so much the more visible; for no root could develop if it were not for what is mediated between the root and the earth-realm by these remarkable root-spirits, which bring the mineral element of the earth into flux in order to conduct it to the roots of the plants. Naturally I refer to the underlying *energetic* process.

These root-spirits, which are everywhere present in the earth, get a quite particular sense of well being from rocks and from ores (which may be more or less transparent). But they enjoy their greatest sense of well-being, because here they are really at home, when they are conveying what is mineral to the roots of the plants. And they are completely enfilled with an inner element of spirituality which we can only compare with the inner element of spirituality in the human eye, in the human ear. For these root-spirits are in their spirit-nature entirely sense. Apart from this they are nothing at all; they consist only of sense. They are entirely sense, and it is a sense which is at the same time understanding, which does not only see and hear, but immediately understands what is seen and heard, which in receiving impressions, receives also ideas.

We can even indicate the way in which these root-spirits receive their ideas. We see a plant sprouting out of the earth. The plant comes, as I shall presently show you, into connection with the extraterrestrial universe; and, particularly at certain seasons of the year, spirit-currents flow from above, *via the Cosmic Substance*, from the blossom and the fruit of the plant down into the roots below, streaming into the earth. And just as we turn our eyes towards the light and see, so do the root-spirits turn their faculty of perception towards what seeps downwards from above, through the plant into the earth. What seeps down towards the root-spirits, that is something which the *Astral* light has sent into the blossoms, which the sun's warmth has sent into the plants, which the air has produced in the leaves, which the distant stars have brought about in the plant's structures. The plant gathers the secrets of the universe, *as the Ideal*

31

Plant and sinks them into the ground, and the gnomes take these secrets into themselves from what seeps down spiritually to them through the plants. And because the gnomes, particularly from autumn on and through the winter, in their wanderings through ore and rock bear with them what has filtered down to them through the plants, they become those beings within the earth which, as they wander, carry the ideas of the whole universe streaming throughout the earth. We look forth into the wide world. The world is built from universal spirit; it is an embodiment of universal *Cosmic Force* ideas, of universal spirit. The gnomes receive, *them via what the Cosmic Substance has to give,* through the plants, which to them are the same as rays of light are to us, the ideas of the universe, and within the earth carry them in full consciousness from metal to metal, from rock to rock.

We gaze down into the depths of the earth not to seek there below for abstract ideas about some kind of mechanical laws of nature, but to behold the roving, wandering gnomes, which are the light-filled preservers of world-understanding within the earth.

Because these gnomes have immediate understanding of what they see, their knowledge is actually of a similar nature to that of man. They are the compendium of understanding, they are entirely understanding. Everything about them is understanding, an understanding however, which is universal, and which really looks down upon human understanding as something incomplete. The gnomes laugh us to scorn on account of the groping, struggling understanding with which we manage to grasp one thing or another, whereas they have no need at all to make use of thought. They have direct perception of what is comprehensible in the world; and they are particularly ironical when they notice the efforts people have to make to come to this or that conclusion. Why should they do this? say the gnomes — why ever should people give themselves so much trouble to think things over? We know everything we look at. People are so stupid — say the gnomes — for they must first think things over.

And I must say that the gnomes become ironical to the point of ill manners if one speaks to them of logic. For why ever should people need such a superfluous thing — a training in thinking? The thoughts are already there. The ideas flow through the plants. Why don't people stick their noses as deep into the earth as the plant's roots, and let what the sun says to the plants trickle down into their noses? Then they would know something! But with logic — so say the gnomes — there one can only have odd bits and pieces of knowledge.

Thus the gnomes, inside the earth, are actually the bearers of the *Spirit's* ideas of the universe, of the world-all. But for the earth itself they have no liking at all. They bustle about in the earth with ideas of the universe, *within the cosmic forces,* but they actually hate what is earthly *substance.* This is something from which the gnomes would best like to tear themselves free. Nevertheless they remain with the earthly — you will soon see why this is — but they hate it, for the earthly *substance* threatens them with a continual danger. The earth continually holds over them the threat of forcing them to take on a particular form, the form of those creatures I described to you in the last lecture, the amphibians, and in particular of the frogs and the toads. The feeling of the gnomes within the earth is really this: If we grow too strongly together with the earth, we shall assume the form of frogs or toads. They are continually on the alert to avoid being caught in a too strong connection with the *earthly substance,* to avoid taking on earthly form. They are always on the defensive against this earthly form, which threatens them as it does because of the element in which they exist. They have their home in the earthly-moist element; there they live under the constant threat of being forced into amphibian forms. From this they continually tear themselves free, by filling themselves entirely with *Cosmic Force* ideas of the extra-terrestrial universe. The gnomes are really that element within the earth which represents the extra -terrestrial, because they must continually reject a growing together with the earthly *substance;* otherwise, as single beings, they would take on the forms of the amphibian world. And it is just from what I may call this feeling of hatred, this feeling of antipathy towards the earthly, that the gnomes gain the *Cosmic Force's* power of driving the plants up out of the earth. With the fundamental force of their being they unceasingly thrust away the earthly, and it is this thrusting that determines the upward direction of the plant's growth; they push the plants up with them. It accords with the nature of the gnomes in regard to the earthly to allow the plant to have only its roots in the earth, and then to grow upwards out of the earth-sphere; so that it is actually out of the force of their own original nature that the gnomes push the plants out of the earth and make them grow upwards. (2)

Crystallisation

The mineral substances must emancipate themselves from what is working immediately above the surface of the Earth, if they wish to be exposed to the most distant cosmic forces. And in our cosmic age they can most easily do so — they can most easily emancipate themselves from the Earth's immediate neighbourhood and come under the influence of the most distant cosmic forces down inside the Earth —in the time between the **15th January and the 15th February**; in this winter season. The time will come when such things are recognised as exact indications. This is the season when the strongest formative-forces of crystallisation, the strongest forces of form, can be developed for the mineral substances within the Earth. It is in the middle of the winter. The interior of the Earth then has the property of being least dependent on itself — on its own mineral masses; it comes under the influence of the crystal-forming forces that are there in the wide spaces of the Cosmos.

This then is the situation. Towards the end of January the mineral substances of the Earth have the greatest longing to become crystalline, and the deeper we go into the Earth, the more they have this longing to become purely crystalline within the "household of Nature." In relation to plant growth, what happens in the minerals at this time is most of all indifferent, or neutral. That is to say, the plants at this time are most left to themselves within the Earth; they are least exposed to the mineral substances. On the other hand, for a certain time before and after

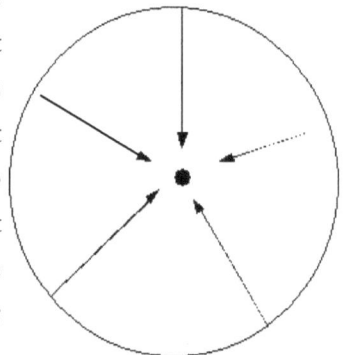

this period — and notably before it, when the minerals are, so to speak, just on the point of passing over into the crystalline element of form and shape — then they are of the greatest importance; they ray out the forces that are particularly important for plant-growth.

Thus we may say, approximately in the month of November-December, there is a point of time when that which is under the surface of the Earth becomes especially effective for plant-growth. The practical question is: "How can we really make use of this for the growth of plants?" The time will come when it is recognised, how very important it is to make use of these facts, so as to be able to direct the growth of plants. I will observe at once, if we are dealing with a soil which does not readily or of its own accord carry upward the influences which should be working upward in this winter season, then it is well to add a dose of clay to the soil. (I shall indicate the proper dose later on). We thereby prepare the soil to carry *the Cosmic Forces* upward, which to begin with, *are* inside the Earth and make it effective for the growth of plants. I mean, the crystalline forces which we observe already when we look out over the crystallising snow. (The force of crystallisation, however, grows stronger and more intense the farther we go into the interior of the Earth). This crystallising force must therefore be carried upward at a time when it has not yet reached its culminating point — which it will only attain in January or February.

Thus we derive the most positive hints from knowledge which at first sight seems remote. We get indications that will really help us, where we should otherwise be experimenting in the dark.

Altogether, we should be clear that the whole domain of Agriculture — including what is beneath the surface of the Earth — represents an individuality, a living organism, living even in time. The life of the Earth is especially strong during the winter season, whereas in summer-time it tends in a certain sense to die. (3)

Wood Sap

With the help of the Gnomes and the Cosmic Forces the plant grows out of the ground. The root grows out of the seed. Let us first take a tree; we can then pass to the ordinary plants. We take a tree: the stem grows up. This growth is very remarkable. This stem which grows there, is really only formed because it lets sap mount from the earth, and this sap in mounting carries up with it all kinds of salts and particles of earth*ly substance*; and so the stem becomes hard. When you look at the wood from the stem of a tree, you have a mounting sap, and this sap carries with it fine particles of earth, and all sorts of salts too, for instance, carbonate of soda, iron, etc., into the plants and this makes hard wood. The essential thing is that the sap mounts.

What happens, in reality? The earthy, the solid, becomes fluid! And we have an earthy-fluid substance mounting there. Then the fluid evaporates and the solid remains behind: that is the wood.

You see, this sap which mounts up in the tree — let us call it wood-sap — is not created there but is already contained everywhere in the earth, so that the earth in this respect is really a great living Being, *with its own Etheric, Astral and Spirit activities.* This sap which mounts in the tree, is really present in the whole earth: only in the earth it is something special. It becomes in the tree what we see there. In the earth it is in fact the *etherized* sap, *working through the Earthly Substance,* which actually gives it life. For the earth is really a living Being; and that which mounts in the tree is in the whole earth and through it the earth lives. In the tree it loses its life-giving quality; it becomes merely a chemical; it has only chemical qualities.

So when you look at a tree, you must say to yourself: the earthy-fluidic in the tree — that has become chemical; underneath in the earth it was still alive. So the wood-sap has partly died, as it mounted up in the tree. Were this all, never would a plant come into existence, but only stumps, dying at the top, in which chemical processes are at work. (1)

Once the plant has grown upwards, once it has left the domain of the gnomes and has passed out of the sphere of the moist-earthly element into the *atmospheric Etheric* sphere of the moist-airy *Earthly Forces*, the plant develops what comes to outer physical formation in the leaves. But in all that is now active in the leaves other beings are at work, water-spirits, elemental spirits of the watery element, to which an earlier instinctive clairvoyance gave among others the name of undines. Just as we find the roots busied about, woven-about by the gnome-beings in the vicinity of the ground, and observe with pleasure the upward-striving direction which they give, we now see these *Etheric* water-beings, these elemental beings of the water, these undines in their connection with the leaves. (2)

Say we plant the seed of some plant in the Earth. Here in this seed we have the stamp or impress of the whole Cosmos — from one cosmic aspect or another, *carried in the Astral sphere by the Saturn I processes.* The constellation takes effect in the seed; thereby it receives its special *archetypal* form. Now, the moment it is planted in the Earth-realm, the external forces of the Earth influence it very strongly, and it is permeated every moment with a longing to deny the cosmic process — that is to say, to grow hypertrophied, to grow out in all manner of directions. For that which is working as *Earthly Forces,* above the Earth does not really want to preserve this form.

Humus

The seed must be driven to the state of chaos, *at pollination.* On the other hand, when the first beginnings of the plant are unfolding out of the seed, *at germination,* and at the later stages also — over against the cosmic form which is living as the plant-form in the seed we need to bring the earthly *forces* element into the plant. We must bring the plant nearer to the Earth in its growth. And this we can only do by bringing into the life of the plant such life as is already present on the Earth. That is to say, we must bring into it life that has not yet reached the utterly chaotic state — life that has not yet gone forward to the stage of seed-formation — life, that is to say, which came to an end in the organisation of some plant before it reached the point of seed-formation.

In effect, we must bring into it such life as is already present on the Earth. In this respect, in districts which are well favoured by fortune, a rich humus-formation comes very largely to man's assistance in "Nature's household." For in the last resort man can but sparingly replace by artificial means the fertility the Earth itself is able to achieve by natural humus-formation. To what is this transformation due? It is due to the fact that that which comes from the plant-life is absorbed by the whole

Nature-process. To some extent, all life that has not yet reached the state of chaos rejects the cosmic influences. If such life is also made use of in the plant's growth, the effect is to hold fast in the plant what is essentially earthly *forces*. The cosmic *force* process works only in the stream which passes upward once more to the seed-formation; while on the other hand the earthly *force* process works in the unfolding of leaf, blossom and so on, and the cosmic only radiates its influences into all this.

The Plant

We can trace the process quite exactly. Assume you have a plant growing upward from the root. At the end of the stem the little grain of seed is formed. The leaves and flowers spread themselves out. Now the earthly *force* element in leaf and flower is the shape and form and the filling of earthly matter. The reason why a leaf or grain develops thick and strong — absorbs inner substantialities, and so on — the reason for this lies in all that which we bring to the plant by way of *earthly* life that has not yet reached the state of chaos. On the other hand, the seed which evolves its *cosmic* force right up the steam (in a vertical direction, not in the circling round) — the seed irradiates the leaf and blossom of the plant with the force of the Cosmos.

We can see this directly. Look at the green plant-leaves. The green leaves, in their form and thickness and in their greeness too, carry an earthly *force* element, but they would not be green unless the cosmic force of the Sun were also living in them. And even more so when you come to the coloured flower; therein are living not only the cosmic forces of the Sun, but also the supplementary forces which the Sun-forces receive from the distant planets — Mars, Jupiter and Saturn, *as Cosmic Substance*. In this way we must look at all plant growth. Then, when we contemplate the rose, in its red colour we shall see the forces of Mars. Or when we look at the yellow sunflower — it is not quite rightly so called, it is called so on account of its form; as to its yellowness it should really be named the Jupiter-flower. For the force of Jupiter, supplementing the cosmic force of the Sun, brings forth the white or yellow colour in the flowers. And when we approach the chicory (Cichoriuns Intybus), we shall divine in the bluish colour the influence of Saturn, supplementing that of the Sun. Thus we can recognise Mars in the red flower, Jupiter in the yellow or white, Saturn in the blue, while in the green leaf we see essentially the Sun itself. But that which thus shines out in the colouring of the flower works as a *cosmic force* most strongly in the root. For the forces that live and abound in the distant planets are working, as we have seen, down there below within the earthly soil.

It is so indeed. We must say to ourselves: Suppose we pull a plant out of the Earth. Down below we have the root. In the root there is the cosmic *force* nature, whereas in the flower most of all there is the earthly *force*, the cosmic *substance* being only present in the delicate quality of the colouring and shading. If on the other hand the earthly *substance* nature is to live strongly in the root, then it must shoot into form. For the plant always has its form from that which can arise within the earthly realm. That which expands the form is *the etheric inspired* earthly *processes*. Thus if the root is ramified and much-divided, then, as in the flower's colouring the cosmic nature is working upward, *stimulating the Cosmic Substance,* so here the earthly nature is working downward, *stimulating the Earthly Substance*. Therefore the cosmic roots are those that are more or less single in form, whereas in highly ramified roots we have a working of the earthly *forces* nature downward into the soil, just as in colour we have a working-upward of the cosmic *forces* nature into the flower.

The Sun-quality is in the midst between the two. The Sun-nature lives most of all in the green leaf, in the mutual interplay between the flower and the root and all that is between them. The Sun-quality is really that which is related, as a "diaphragm" (for so we called it in picture 9) with the surface of the earth. The cosmic *forces are* associated with the interior of the Earth and works upward into the upper parts of the plant, *interacting with the Cosmic Substance*. The earthly *forces*, which *are* localised above the surface of the earth, works downward, being carried down into the plant with the help of the limestone element, *to stimulate the Earthly Substance*.

Observe those plants in which the limestone strongly draws the earthly *force* nature downward into the roots. These are the plants whose roots shoot out in all directions with many ramifications, such, for instance, as the food fodder plants — I do not mean turnips or the like, but plants like sainfoin. Such things must be recognised in the form of the plant. To understand the plant, we must recognise the form of the plant and from the colour of the flower, the extent to which the cosmic and the earthly *processes* are working there.

Assume that by some means we cause the cosmic *forces* to be strongly retained — held up within the plant itself. Then it will not reveal itself to any great extent. It will not shoot out into blossom but will express itself in a stalk -like nature. Where, now, according to the indications we have given, does the cosmic nature live in the plant? It lives in the silicious element.

Look at the equisetum plant. It has this peculiarity: it draws the cosmic nature to itself; it permeates itself with the silicious nature. It contains no less than 90% of silicic acid. In equisetum the cosmic *force* is present, so to speak, in very great excess, yet in such a way that it does not go upward and reveal itself in the flower but betrays its presence in the growth of the lower parts.

Or let us take another case. Suppose that we wish to hold back *the cosmic forces* in the root-nature of a plant that which would otherwise tend upward through the stem and leaf. No doubt this is not so important in our present earthly epoch, for through various conditions we have already largely fixed the different species of plants. In former epochs — notably in primeval epochs — it was different. At that time it was still possible quite easily to transform one plant into another; hence it was very important to know these things. To-day too, it is important if we wish to find what conditions are favourable to one plant or another.

What do we then need to consider? How must we look at a plant when we desire the cosmic forces not to shoot upward into the blossoming and fruiting process but to remain below? Suppose we want the stem and leaf-formation to be held back in the root. What must we then do? We must put such a plant into a sandy soil, for in silicious soil—*especially with high phosphorus levels* – the cosmic is held back; it is actually "caught:" Take the potato, for example. With the potato this end must be attained. The

blossoming process must be kept below. For the potato is a stem and leaf
-formation down in the region of the root. The leaf and stem-forming
process is held back, retained in the potato itself. The potato is not a root,
it is a stem-formation held back. We must therefore bring it into a sandy
soil. Otherwise we shall not succeed in having the cosmic force retained
in the potato.

This, therefore, is the ABC for our judgment of plant-growth. We must
always be able to say, what in the plant is cosmic, and what is terrestrial or
earthly. How can we adapt the soil of the earth, by its special consistency,
as it were to densify the cosmic *forces* and thereby hold it back more in the
root and leaf? Or again, how can we thin it out so that it is drawn upward
in a dilute condition, right up into the flowers, *connecting more strongly with
the Cosmic Substance process,* giving them colour — or into the fruit-forming
process, permeating the fruit with a fine and delicate taste? For if you
have apricots or plums with a fine taste — this taste, just like the colour
of the flowers, is the cosmic *force* quality which has been carried upward,
right into the fruit. In the apple you are eating Jupiter, in the plum you are
actually eating Saturn.

If mankind with their present state of knowledge were suddenly obliged
to create, from the comparatively few plants of the primeval epoch of the
Earth, the manifold variety of our present fruits and fruit-trees, they
would not get very far. We should not get far if it were not for the fact
that the forms of our different fruits are inherited. They were produced at
a time when humanity had knowledge, out of primeval and instinctive
wisdom, how to create the different kinds of fruits from the primitive
varieties that then existed. If we did not already possess the different
kinds of fruit, handing them down by heredity —if we had to do it all
over again with our present cleverness — we should not be very
successful in creating the different kinds of fruit. Nowadays it is all done
by blind experiment, there is no rational penetration into the process.

As I said just now, the man of to-day may know — though this
knowledge too is very scanty — he may know how the air behaves in the
interior of the Earth. But he knows practically nothing of how the light

behaves in the interior of the Earth. He does not know that the silicious — that is, the cosmic *carrier* — stone or rock or sand receives the light into the Earth and makes it effective there. Whereas that which stands nearer to the earthly-living nature, namely the humus, does not receive it; it does not make the light effective in the Earth. It therefore gives rise to a "light-less" working. Such things must be penetrated once more with clear understanding. (3)

But the stem, formed from *the Wood sap, which is a combination of Cosmic forces and Earthly Substances, that have gone through the fructification event,* rises into the air, and the air always contains moisture. It comes into the moist air, it comes with the sap which has created it, from the earthy-fluidic into the fluidic-airy *drenched Earthly Forces* and life springs up in it anew, so that around it green leaves appear and finally flowers. ... Again there is life. You see, in the foliage, in the leaf, in the bud, in the blossom, there is once more the sap of life; the wood-sap is dead life-sap.

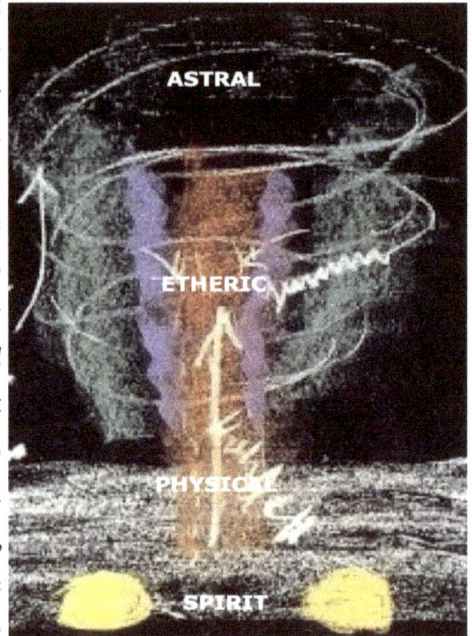

In the stem, life is always dying; in the leaf it is always being resurrected. So that we must say: We have wood sap, which mounts *from the Earth*; then we have life-sap *from the Etheric and Astral activity of the environment.* And what does this do! It travels all round and brings forth the leaves everywhere: so that you can see the spirals in which the leaves are arranged. The living sap really circles round. It arises from the fluid-airy element into which the plant comes when it has grown out of the earthy-fluidic element.

The stem, the woody stem, is dead and only that which sprouts forth around the plant is alive. This you can easily prove in the following very simple way. Go to a tree: you have the stem, then the bark, and in the bark the leaves grow. Now cut the bark away at that point; the leaves come away too. At this point leave the leaves with the bark. The result is that there the tree remains fresh and living, and here it begins to die. The wood alone with its sap cannot keep the tree alive; what comes with the leaves must come from outside and that again contains *etheric filled* life, *from the oxygen in the atmosphere.* We see in this way that the earth can certainly put forth the tree, but she would have to let it die if it did not get *etheric* life from the damp air: for in the tree the sap is only a chemical, no giver of life. The living sap that circulates, that gives it life. And one can really say: When the sap rises in the spring, the tree is created anew; when the living sap again circulates in the spring, every year the tree's life is renewed. The earth produces the *Wood* sap, *with the Spiritualised Cosmic Forces and Physicalised Earthly Substance*, from the earthy-fluidic; the fluidic-airy produces the living sap, *through the action of the Ethericised Earthly Forces and Astralised Cosmic Substance.* (1)

These undine beings differ in their inner nature from the gnomes. They cannot turn like a spiritual sense-organ outwards towards the universe.

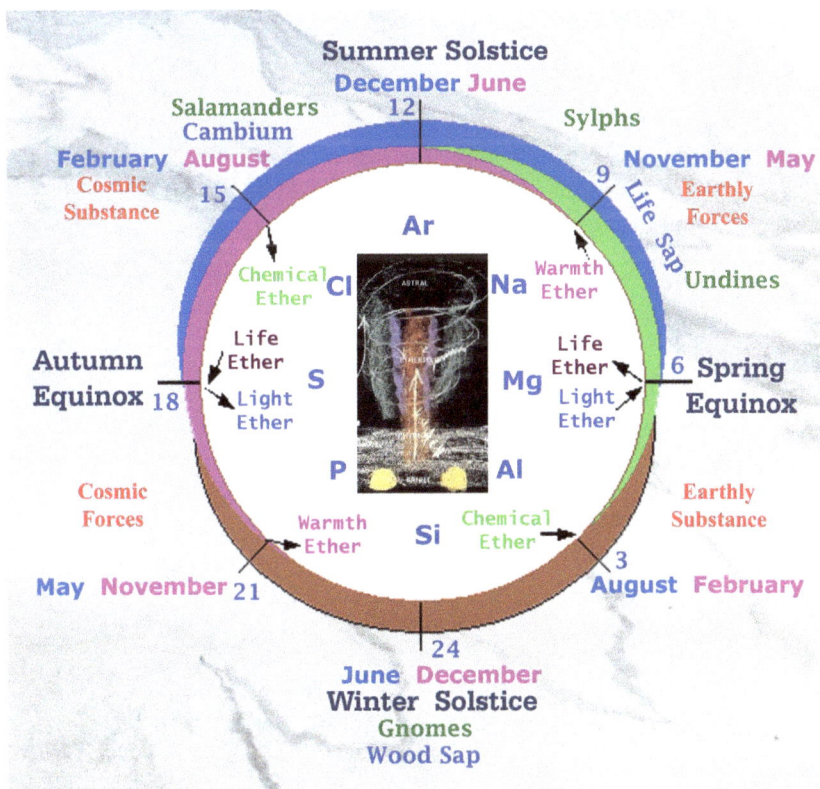

They can only yield themselves up to the weaving and working of the whole cosmos in the airy-moist element, and therefore they are not beings of such clarity as the gnomes. They dream incessantly, these undines, but their dream is at the same time their own form. They do not hate the earth as intensely as do the gnomes, but they have a sensitivity to what is earthly. They live in the etheric element of water, swimming and swaying through it, and in a very sensitive way they recoil from everything in the nature of a fish; for the fish-form is a threat to them, even if they do assume it from time to time, though only to forsake it immediately in order to take on another metamorphosis. They dream their own existence. And in dreaming their own existence they bind and release, they bind and disperse the substances of the air, *with the help of the Earthly Forces,* which in a mysterious way they introduce into the leaves,

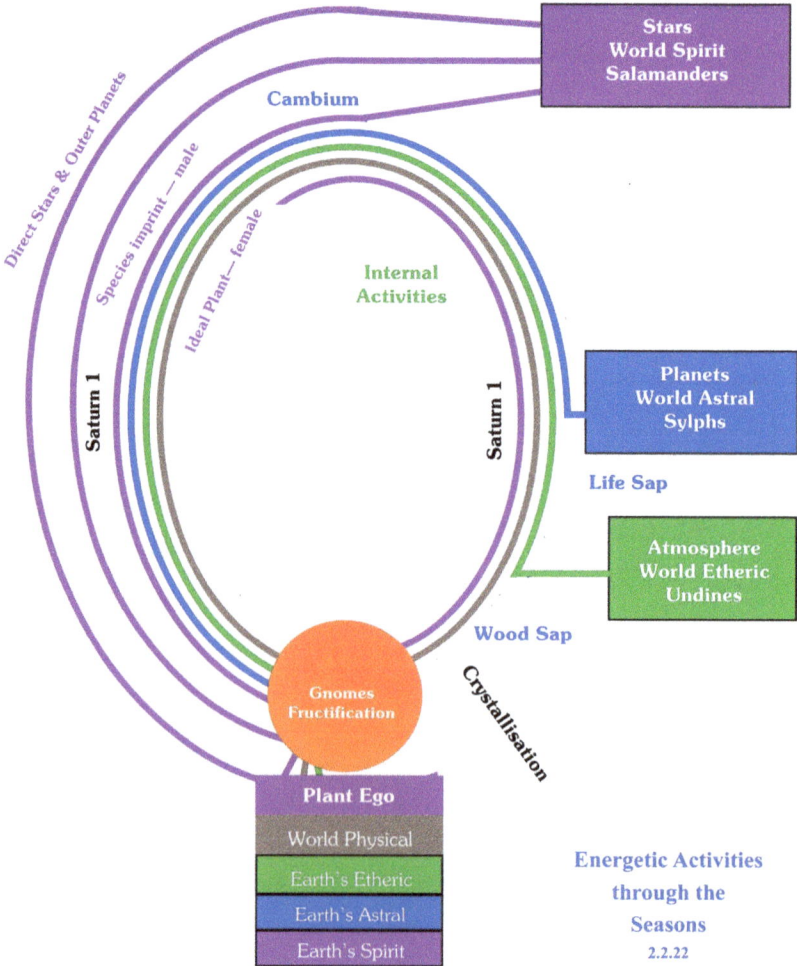

Stars
World Spirit
Salamanders

Cambium

Direct Stars & Outer Planets

Species imprint — male

Ideal Plant — female

Internal
Activities

Saturn 1

Saturn 1

Planets
World Astral
Sylphs

Life Sap

Atmosphere
World Etheric
Undines

Wood Sap

Crystallisation

Gnomes
Fructification

Plant Ego

World Physical

Earth's Etheric

Earth's Astral

Earth's Spirit

Energetic Activities
through the
Seasons
2.2.22

as these are pushed upwards by the gnomes. For at this point the plants would wither if it were not for the undines, who approach from all sides, and show themselves, as they weave around the plants in their dream-like existence, to be what we can only call the world-chemists. The undines dream the uniting and dispersing of substances. And this dream, in which the plant has its existence, into which it grows when, developing upwards, it forsakes the ground, this undine-dream is the world-chemist which brings about in the plant-world the mysterious combining and separation of the substances which emanate from the leaf. We can therefore say that the undines are the chemists of plant-life. They dream of chemistry. They possess an exceptionally delicate spirituality which is really in its element just where *etheric* water and *astral* air come into contact with each other. The undines live entirely in the element of moisture, but they develop their actual inner function when they come to the surface of something watery, be it only to the surface of a water-drop or something else of a watery nature. For their whole endeavour lies in preserving themselves from getting the form of a fish, the permanent form of a fish. They wish to remain in a condition of metamorphosis, in a condition of eternal, endlessly changing transformation. But in this state of transformation in which they dream of the stars and of the sun, of light and of warmth, they become the chemists who now, starting from the leaf, carry the plant further in its *etheric* formation, after it has been pushed upwards by the power of the gnomes. So the plant develops its leaf growth, and this mystery is now revealed as the dream of the undines, into which the plants grow.

To the same degree, however, in which the plant grows into the dream of the undines, does it now come into another domain, into the domain of those spirits which live in the airy-warmth element, just as the gnomes live in the moist-earthly *substance,* and the undines in the moist-airy *earthly forces* element. Thus it is in the *cosmic substance* element which is of the nature of air and warmth that those beings live which an earlier clairvoyant art designated as the sylphs. Because air is everywhere imbued with *Astral* light, these sylphs, which live in the airy-warmth element, press towards the light, relate themselves to it. They are particularly susceptible to the finer but larger movements within the atmosphere.

When in spring or autumn you see a flock of swallows, which produce as they fly vibrations in a body of air, setting an air-current in motion, then this moving air-current — and this holds good for every bird — is for the sylphs something audible. Cosmic music sounds from it to the sylphs. If, let us say, you are travelling somewhere by ship and the seagulls are flying around it, then in what is set in motion by the seagulls' flight there is a spiritual sounding, a spiritual music which accompanies the ship.

Again it is the sylphs which unfold and develop their being within this sounding music, finding their dwelling-place in the moving current of air. It is in this spiritually sounding, moving element of air that they find themselves at home; and at the same time they absorb what the power of *Astral* light sends into these vibrations of the air. Because of this the sylphs, which experience their existence more or less in a state of sleep, feel most in their element, most at home, where birds are winging through the air. If a sylph is obliged to move and weave through air devoid of birds, it feels as though it had lost itself. But at the sight of a bird in the air something quite special comes over the sylph. I have often had to describe a certain event in man's life, that event which leads the human soul to address itself as "I". And I have always drawn attention to a saying of Jean Paul, that, when for the first time a human being arrives at the conception of his "I", it is as though he looks into the most deeply veiled Holy of Holies of his soul. A sylph does not look into any such veiled Holy of Holies of its own soul, but when it sees a bird an ego-feeling comes over it. It is in what the bird sets in motion as it flies through the air that the sylph feels its ego, *or its internal spirit*. And because this is so, because its ego is kindled in it from outside, the sylph becomes the bearer of cosmic love through the atmosphere. It is because the sylph embodies something like a human wish, but does not have its ego within itself but in the bird-kingdom, that it is at the same time the bearer of wishes of love through the universe.

Thus we behold the deepest sympathy between the sylphs and the bird-world. Whereas the gnome hates the amphibian world, whereas the undine is unpleasantly sensitive to fishes, is unwilling to approach them, tries to avoid them, feels a kind of horror for them, the sylph, on the other hand, is attracted towards birds, and has a sense of well-being when it can waft towards their plumage the swaying, love-filled waves of the air. And were you to ask a bird from whom it learns to sing, you would hear that its inspirer is the sylph. Sylphs feel a sense of pleasure in the bird's form. They are, however, prevented by the cosmic ordering from becoming birds, for they have another task. Their task is lovingly to convey *Astral* light to the plant. And just as the undine is the *Etheric* chemist for the plant, *stimulating the Earthly Forces,* so is the sylph the *Astral* light-bearer *using the Cosmic Substance process*. The sylph imbues the plant with light; it bears light into the plant.

Through the fact that the sylphs bear light and *Cosmic Substances* into the plant, something quite remarkable is brought about in it. You see, the sylph is continually carrying light into the plant. The *Cosmic Substance* light, that is to say the power of the sylphs in the plant, works upon the chemical forces *working with the Earthly Forces,* which were induced into the plant by the undines. Here occurs the inter-working of sylph-light and undine-chemistry. This is a remarkable plastic activity. With the help of the upstreaming substances, *provided through the combined activity of the Cosmic Forces and Earthly Substances,* which are worked upon by the undines, the sylphs weave out of the light an ideal plant-form. They actually weave the *Ideal* Plant within the plant from *Cosmic Substance* light, *from the Sylphs* and from the chemical *Earthly Force* working of the undines. And when towards autumn the plant withers and everything of physical substance disintegrates, then these *'female'* plant-forms begin to seep downwards, and now the gnomes perceive them, perceive what the world — the sun through the sylphs, the air through the undines — has brought to pass in the plant. This the gnomes perceive, so that throughout the entire winter they are engaged in perceiving below what has seeped into the ground through the plants. Down there they grasp world-ideas in the plant-forms which have been plastically developed with the help of the sylphs, and which now in their spiritual *feminine* ideal form enter into the ground.

Naturally those people who regard the plant as something purely material know nothing of this spiritual *feminine* ideal form. Thus at this point something appears which in the materialistic observation of the plant gives rise to what is nothing other than a colossal error, a terrible error. I will sketch this error for you.

Fructification

Everywhere you will find that materialistic science describes matters as follows: The plant takes root in the ground, above the ground it develops its leaves, finally unfolding its blossoms, within the blossoms the stamens, then the seed-bud. Now — usually from another plant — the pollen from the anthers, from the pollen vessels, is carried over to the germ which is then fructified, and through this the seed of the new plant is produced. The germ is regarded as the female element and what comes from the stamens as the male — indeed matters cannot be regarded otherwise as long as people remain fixed in materialism, for then this process really does look like a fructification.

This, however, it is not. In order to gain insight into the process of fructification, that is to say the process of reproduction, in the plant-world, we must be conscious that in the first place it is from what the great chemists, the undines, bring about in the plants, and from what the sylphs bring about, that the *female* plant-form arises, *as the best expression of what that plant can be in that seasons conditions. This is the female* ideal plant-form which sinks into the ground and is preserved by the gnomes. It is there below, this plant-form. And there within the earth it is now guarded by the gnomes after they have seen it, after they have looked upon it. The earth becomes the mother-womb for what thus seeps downwards. This is something quite different from what is described by materialistic science.

The Ideal Plant — Cosmic Substance , Astral stream

After it has passed through the sphere of the sylphs, the plant comes into the sphere of the *Spirit filled* elemental fire-spirits. These fire-spirits are the inhabitants of the warmth-light element. When the warmth of the earth is at its height, or is otherwise suitable, they gather the *Spirits* warmth together, *within the Cosmic Substance process.* Just as the sylphs gather up the light, so do the fire-spirits gather up the warmth and carry it into the blossoms of the plants. (2)

But that is not all. While this is happening, between the bark, still full of living sap, and the woody stem, there is formed a new layer. Now I cannot say that a sap is formed. I have already spoken of wood-sap, living sap, but I cannot again say that a sap is formed: for what is formed is quite solid: it is called cambium. It is formed between the bark which still belongs to the leaves, and the wood. But the plant needs cambium too, in a certain way. You see, the wood sap is formed in the earthy-fluidic, the life sap in the fluidic-airy, and the cambium in the warm air, in the warm damp, or the airy-warmth *of the astralised Cosmic Substance process.* The plant develops warmth while it takes up life from outside, *through the interaction between the Cosmic Substance and Earthly Forces, which must reach a certain intensity. As indicated by the swirling nature of the image on page 42.* This warmth goes inward and develops the cambium inside. Or if the cambium does not yet develop — the plant needs cambium and you will shortly hear why — before the cambium forms, there is first of all developed a thicker substance: the plant gum. Plants form this plant gum in their inner warmth, and this, under certain conditions, is a powerful means of healing. Thus the *Wood* sap carries the plant upwards, the leaves give the plant *etheric* life, then the leaves by their warmth produce the gum which reacts on the warmth. And in old plants, this gum, running down to the ground, has become transparent. When the earth was less dense and damper, the gum became transparent and turned to Amber. You see, then, when you take up a piece of Amber, what from prehistoric plants ran down to the ground as resin and pitch. This the plant gives back to the earth: Pitch, Resin, Amber. And if the plant retains it, it becomes cambium. Through the *Wood* sap the plant is connected with the earth; the life-sap brings the plant into connection with what circulates round the earth — with the airy-moist circumference of the earth. But *the Cosmic Substance activity of* the cambium brings the plant into connection with the stars, with what is above, and in such a way that within this cambium the *female ideal* form of the next plant develops. This passes over to the seeds and in this way the next plant is born, so that the stars indirectly through *the Cosmic Substance* and the cambium create the next plant! So that the plant is not merely created from the seed — that is to say, naturally it is created from the seed, but the seed must first be worked on by *the Cosmic Substance process and* the cambium, that is: by the whole heavens.

It is really wonderful — a seed, a humble, modest little seed could only come into existence because the cambium — now not in liquid but in solid form — imitates the whole plant; and this form which arises there in the cambium — a new *female* plant form — this carries the *reproductive* power to the seed to develop through the forces of the earth into a new plant.

Through mere speculation, when one simply puts the seed under a microscope, nothing is gained. We must be clear what parts the *wood* sap, the life sap, the cambium, play in the whole matter. The wood sap is a relatively thin sap: it is peculiarly fitted to allow chemical changes to take place in it. The life sap is certainly much thicker, it separates off its gum. If you make the gum rather thick, you can make wonderful figures with it. Thus the life sap, more pliable than the wood sap, clings more to the plant -form. And then it gives this up entirely to the cambium. That is still thicker, indeed quite sticky, but still fluid enough to take the forms which are given it by the stars, *via the Cosmic Substance.*

So it is with trees, and so, too, with the ordinary plants. When the rootlet is in the earth, the sprout shoots upward. But it does not separate off the solid matter, does not make wood; it remains like a cabbage stalk. The leaves come out directly on the circumference, in spirals, the cambium is formed directly in the interior, and the cambium takes everything back to the earth with it. So that in the annual plants the whole process occurs much more quickly. In the tree, only the hard parts are separated out, and not everything is destroyed.

The same process occurs in ordinary plants too, but is not carried so far as in trees. In the tree it is a fairly complicated matter. When you look at the tree from above, you have first the pith inside: this gives the direction. Then layers of wood form round the pith. Towards the autumn the gum appears from the other side, and fastens the layers together. So we have the gummy wood of one year. In the next year this is repeated. Wood forms somewhere else, is again gummed together in the autumn, and so the yearly rings are formed. So you see everything clearly if only you understand that there are three things: wood sap, life sap, and cambium. The wood sap is the most fluid, it is really a chemical; the life sap is the giver of life; it is really, if I may so express myself, a living thing. And as for the cambium, there the whole plant is sketched out *from what the Cosmic Substance receives from* the stars. It is really so. The wood sap rises and dies, then life again arises; and now comes the influence of the stars, so that from the thick, sticky cambium the new *'female'* plant is sketched out. In the cambium one has a sketch, a sculptural activity. The stars model in it from the whole universe the complete plant form. So you see, we come from Life into the Spirit. What is modelled there is modelled from out of the World-Spirit. The earth first gives up her life to the plant, the plant dies, the air environment along with its light once more gives it life, and the World Spirit implants the new plant form. This is preserved in the seed and grows again in the same way. So that one sees in the growing plant how the plant world rises out of the earth, through death, to the living Spirit.

Wood sap develops in man as the ordinary colourless mucus. Wood sap in plants is, in man, mucus. The life sap of the plant which circulates from the leaves, corresponds to the human blood. And the cambium of the plant corresponds to the milk and the chyle in the human being. When a woman begins to nurse, certain glands in the breast cause a greater flow of milk. Here you have again something in human beings which is most strongly influenced by the stars, namely, *milk.* Milk is absolutely necessary for the development of the brain — the brain, one might almost say, is solidified milk. Decaying leaves create no proper cambium because they no longer have the power to work back into the proper warmth. They let the warmth escape outwards from the dying edges instead of sending it inwards. We eat these plants with an improperly developed cambium: they do not develop a proper milk; the women do not produce proper milk; the children get milk on which the stars cannot work strongly, and therefore they cannot develop properly. (1)

Cosmic Male Seed — Cosmic Forces, Spirit stream

Now for the tilling of the soil one important thing should above all be understood. We must know the conditions under which the cosmic spaces are able to pour their forces down into the earthly realm. To recognise these conditions, let us take our start from the seed-forming process. The seed, out of which the embryo develops, is usually regarded as a very complicated molecular structure, and scientists are especially anxious to understand it in its complex molecular structure. In simple molecules, they imagine, there is a simple structure; then it grows ever more complicated, till at last we get to the infinitely complex structure of the protein molecule.

With wonder and astonishment they stand before what they imagine as the complicated structure of the protein in the seed. For they conceive it as follows. They think the protein molecule must be extremely complicated; for after all, out of its complexity, the whole new organism will grow. The new organism, infinitely complex as it is, was already pre-figured in the embryonic condition of the seed. Therefore this microscopic or ultra-microscopic substance must also be infinitely complex in its structure.

To begin with, to a certain extent this is quite true. When the earthly protein is built up, the molecular structure is indeed raised to the highest complexity. But a new organism could never arise out of this complexity. The organism does not arise out of the seed in that way at all. That which develops as the seed, out of the mother-plant or mother-animal, does not by any means simply continue its existence in that which afterwards arises as the descendant plant or animal. That is not true. The truth is rather this:—

When the complexity of structure has been enhanced to the highest degree, *by the Cosmic Substance,* it all disintegrates again, and eventually, where we first had the highest complexity attained within the Earth domain, we now have a tiny realm of chaos. It all disintegrates, as we might say, into cosmic dust. Then, when the seed — having been raised to the highest complexity — has fallen asunder into cosmic dust and the tiny realm of chaos is there, then the entire surrounding Universe begins to work and stamps itself upon the seed, thus building up out of the tiny chaos that which can only be built in it by *Cosmic* forces pouring in from the great Universe from all sides. So in the seed *during pollination*, we get an image of the Universe.

In every seed-formation, the earthly process*es* of organisation is carried to the very end — to the point of chaos. Time and again, in the chaos of the seed the new organism is built up again out of the whole Universe. The parent organism has to play this part: through its affinity to a particular cosmic situation, it tends to bring the seed into that situation whereby the *Cosmic* forces work from the right cosmic directions, so that a dandelion brings forth, not a barberry, but a dandelion in its turn.

That which is imaged in the single plant, is always the image of some cosmic constellation. Ever and again, it is built out of the Cosmos. Therefore, if ever we want to make the *Cosmic* forces effective in our earthly realm, we must drive the earthly as far as possible into a state of chaos. For plant-growth, Nature herself will see to it to some extent, that this is done. However, since every new organism is built out of the Cosmos, it is also necessary for us to preserve the cosmic *force* process, *carried by the primary Saturn stream,* in the organism long enough — that is, until the seed-forming process occurs once more, *if we wish to maintain true species expression. (3)*

Undines carry the action of the chemical ether, *and the Earthly Forces* into the plants, sylphs the action of the light-ether *and Cosmic Substance* into the plant's blossoms. And the pollen now provides what may be called little air-ships, to enable the fire-spirits to carry *the Cosmic Force* warmth into the seed. Everywhere warmth is collected with the help of the stamens, and is carried by means of the pollen from the anthers to the seeds and the seed vessels. And what is formed here in the *Cosmic Force drenched* seed-bud is entirely the male element which comes from the cosmos. It is not a case of the seed-vessel being female and the anthers of the stamens being male. In no way does fructification occur in the blossom, but only the pre-forming of the male seed. (2)

The fructifying force is what the fire-spirits in the blossom take from the warmth of the world-all as the cosmic male seed, which is united with the female element. This *female* element, drawn from the forming of the plant has, as I told you, already earlier seeped down into the ground as *the* ideal form *possible from that seasons growth processes,* and is resting there below. For plants the earth is the mother, the heavens the father. And all that takes place outside the domain of the earth is not the mother-womb for the plant. It is a colossal error to believe that the mother-principle of the plant is in the seed-bud. The fact is that this is the male-principle, which *are Cosmic Forces* drawn forth from the universe with the aid of the fire-spirits. The mother comes from the cambium, *via the Cosmic Substance* ,which spreads from the bark to the wood, and is carried down from above as ideal form. And what now results from the combined working of gnome-activity and fire-spirit activity — this is fructification. The gnomes are, in fact, the spiritual midwives of plant-reproduction. Fructification takes place below in the earth during the winter, when the *male* seed comes into the earth and meets with the *female* forms which the gnomes have received from the activities of the sylphs and undines and now carry to where these forms can meet with the fructifying seeds.

You see, because people do not recognize what is spiritual, do not know how gnomes, undines, sylphs and fire-spirits — which were formerly called salamanders — weave and live together with plant-growth, there is complete lack of clarity about the process of fructification in the plant world. There, outside the earth nothing of fructification takes place, but the earth is the mother of the plant-world, the heavens the father. This is the case in a quite literal sense. Plant-fructification takes place through the fact that the gnomes take from the fire-spirits, *the Cosmic Forces* the fire-spirits have carried into the seed bud as concentrated cosmic warmth on the little airships of the anther-pollen. Thus the fire-spirits are the bearers of warmth.

Once Again

And now you will easily gain insight into the whole process of plant-growth. First, with the help of *the Cosmic Forces coming* from the fire-spirits, the gnomes down below instill *the Earths Etheric* life into the plant and push it upwards. They are the fosterers of life. They carry the life-ether *and Earthly*

Substances to the root — the same life-ether in which they themselves live. The undines foster the chemical ether *and the Earthly Forces*, the sylphs the light-ether *and Cosmic Substance*, the fire-spirits the warmth ether *and Cosmic Forces*. And then the fruit of the warmth-ether *and Cosmic Forces* again unites with what is present below as life *ether and Earthly Substance*. Thus the plants can only be understood when they are considered in connection with all that is circling, weaving and living around them, and one only reaches the right interpretation of the most important process in the plant when one penetrates into these things in a spiritual way.

When once this has been understood, it is interesting to look again at that memorandum of Goethe's where, referring to another botanist, he is so terribly annoyed because people speak of the eternal marriage in the case of the plants above the earth. Goethe is affronted by the idea that marriages should be taking place over every meadow. This seemed to him something unnatural. In this Goethe had an instinctive but very true feeling. He could not as yet know the real facts of the matter, nevertheless he instinctively felt that fructification should not take place above in the blossom. Only he did not as yet know what goes on down below under the ground, he did not know that the earth is the mother-womb of the plants. But, that the process which takes place above in the blossom is not what all botanists hold it to be, this is something which Goethe instinctively felt.

You are now aware of the inner connection between plant and earth. But there is something else which you must take into account.

You see, when up above the fire-spirits are circling around the plant and transmitting *the archetype Saturn 1 Cosmic Forces to* the anther-pollen, then they have only one feeling, which they have in an enhanced degree, compared to the feeling of the sylphs. The sylphs experience their self, their ego, when they see the birds flying about. The fire-spirits have this experience, but to an intensified degree, in regard to the butterfly-world, and indeed the insect-world as a whole. And it is these fire-spirits which take the utmost delight in following in the tracks of the insects' flight so that they may bring about the distribution of warmth for the seed buds. In order to carry the concentrated *male Cosmic Force* warmth, which must descend into the earth so that it may be united with the *female* ideal form, in order to do this the fire-spirits feel themselves inwardly related to the butterfly-world, and to the insect-creation in general. Everywhere they follow in the tracks of the insects as they buzz from blossom to blossom. And so one really has the feeling, when following the flight of insects, that each of these insects as it buzzes from blossom to blossom, has a quite special aura which cannot be entirely explained from the insect itself. Particularly the luminous, wonderfully radiant, shimmering, aura of bees, as they buzz from blossom to blossom, is unusually difficult to explain. And why? It is because the bee is everywhere accompanied by a fire-spirit which feels so closely related to it that, for spiritual vision, the bee is surrounded by an aura which is actually a fire-spirit. When a bee flies through the air from plant to plant, from tree to tree, it flies with an aura which is actually given to it by a fire-spirit. The fire-spirit does not only gain a feeling of its ego in the presence of the insect, but it wishes to be completely united with the insect.

Through this, however, insects also obtain that power about which I have spoken to you, and which shows itself in a shimmering forth of light into the cosmos. They obtain the power completely to spiritualize the physical matter which unites itself with them, and to allow the spiritualized physical substance to ray out into cosmic space. But just as with a flame it is the warmth in the first place which causes the light to shine, so, above the surface of the earth, when the insects shimmer forth into cosmic space, it is the fire spirits which inspire the insects to this activity, the fire-spirits which are circling and weaving around them. But if the fire-spirits are active in promoting the outstreaming of spiritualized matter into the cosmos, they are no less actively engaged in seeing to it that the concentrated fiery element, the concentrated warmth, goes into the interior of the earth, *as Cosmic Forces* so that, with the help of the gnomes, the spirit-form, which sylphs and undines cause to seep down into the earth, may be awakened *through Fructification, to form the Wood Sap.* (2)

Bibliography

(1) Cosmic Workings in Earth and Man Lecture 5, 31 October 1923

(2) Man as Symphony of the Creative Word - Lecture 7, 2nd November, 1923

(3) Agriculture Course Lecture 2, 10 June 1924 (GA 327)

(4) Lecture 6 GA 327

(5) Lecture 8 GA 327

(6) The Groups souls of Animals Plants and Minerals, 2 feb 1908

(7) The Spirit in the realm of plants, 8 dec 1910

(8) Human Questions Cosmic Answers, 2 jul 1922 GA213

(9) The Driving Forces of Spiritual Powers in the World History, 22 Mar 1923 GA 222

(10) Medical lectures 1920

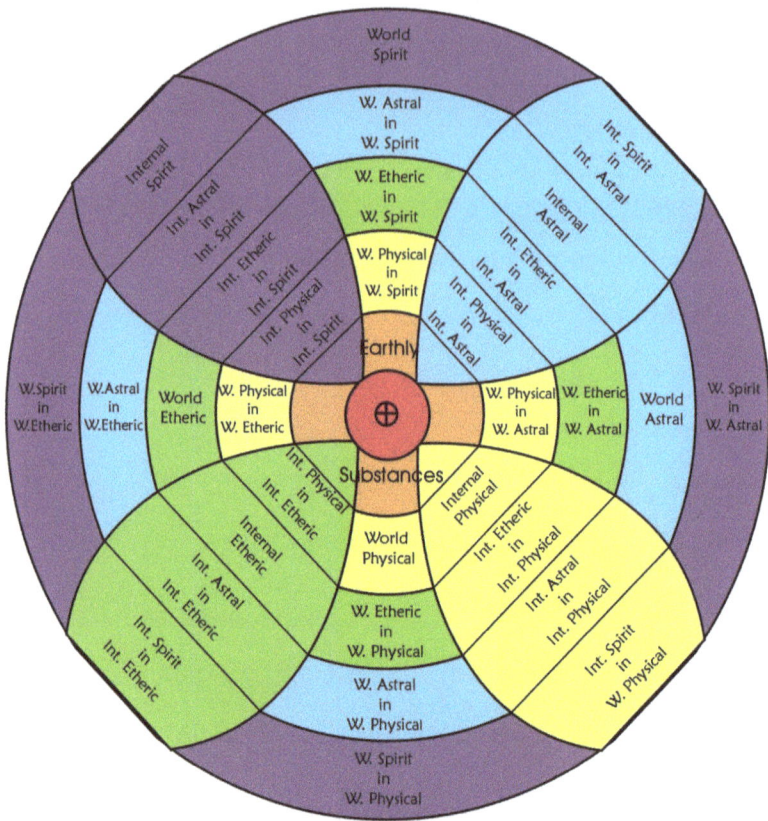

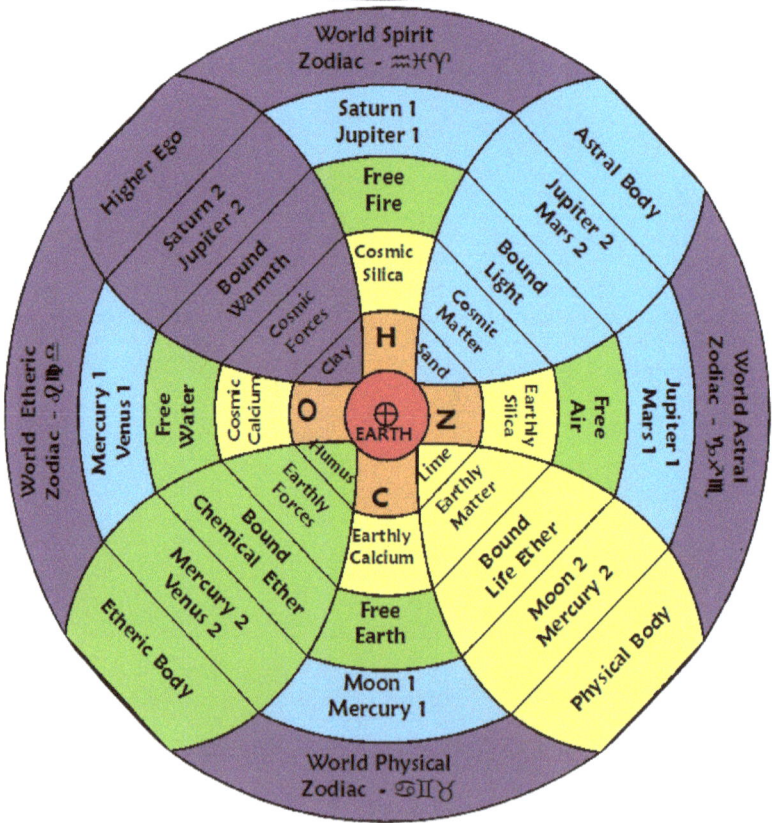

The Energetic Bodies

GA is Italics

We need to look more deeply into the nature of the energetic bodies, and how they appear in living forms. Let us for a moment review the picture of a human being that Anthroposophy gives us. The human being stands before us in a physical body, which has a long evolution behind it, three preparatory stages before it became an earthly body — as is described in my book An Outline of Esoteric Science. This earthly body needs to be understood much more than it is by today's anatomy and physiology. For the human physical body as it is today is a true image of the etheric body, and of the astral body, and even to a certain degree of the Ego organization that humans first received on earth. All of this is stamped like the stamp of a seal upon the physical body — which makes the physical body extraordinarily complicated. Only its purely mineral and physical nature can be understood with the methods of knowledge that are brought to it today. What the etheric body impresses upon it is not to be reached at all by those methods. It has to be observed with the eye of a sculptor so that one obtains pictorial images of cosmic forces, images that can then be recognized again in the form of the entire human being and in the forms of the single organs.

The Physical body is available for us to observe with our scientific mind. The Etheric body, being sourced from our life sustaining Atmosphere, is the energetic body that imbues life and the capacity to grow and develop into all living forms. The World Etheric activity is concentrated by the Earth, with the help of oxygen, water and the alkali elements; and in life forms is seen to move upwards from the Earth, with a seemingly unlimited capacity to create growth. It drives or energises the movement of fluids in general, which leads to the formation of mass and bulk in physical forms. It is especially active in the support the lymphatic and immune system of physical organisms. When the Etheric body leaves a living entity, death occurs. We experience a greater or lesser Etheric activity as a physical sensation of being energetically 'up' or 'down'. When your Etheric body is strong you are full of life. Often a white glow can be seen up to two inches from the physical body when an individuals Etheric body is strong. When it is weak you feel physically down and heavy. The individual with a weak Etheric body looks grey and feels drained.

The Etherics raw activity can be seen in the first stages of growth where there are unshaped forms. For example in the first months of a baby's life or the early stages of plant development. The early leaves of many plants are full and round and repetitive in comparison to the more adult leaves that are often indented and pointed. It produces watery shapes and curved forms.

The mineral kingdom does not undergo changes through a death process, as the Etheric body (and the others) stays external. They remain as atmospheric forces of light heat and water, which work ONTO the mineral externally. The Etheric body is only embodied by plant, animal and human. Crystals may appear to be an exception, however they grow from additions to the outside of their form.

The Plant Kingdom has a physical form and internalises the Etheric body, so it can grow and expand as well as die. With plants the planetary Astral and star sourced Spirit/Ego activities remain external in the Cosmic and World spheres. So, while plants will expand and grow, it is still the external influences of light, moisture and heat that determine their final form. Flowering and seeding are processes bought about by the Astral and Spirit acting onto the plant externally. If they do not work

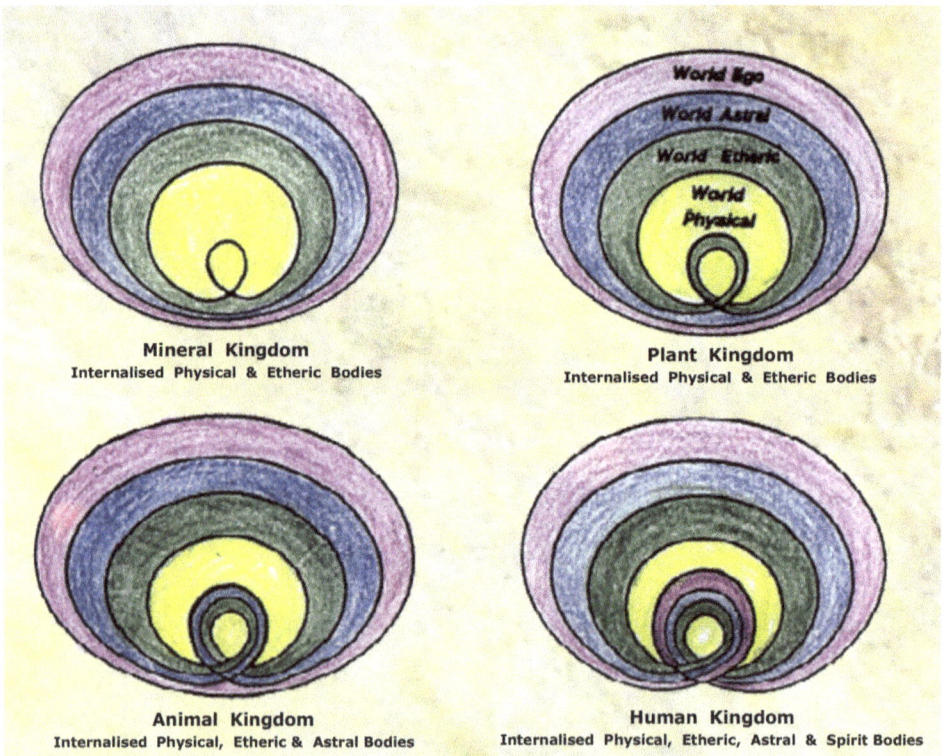

Mineral Kingdom
Internalised Physical & Etheric Bodies

Plant Kingdom
Internalised Physical & Etheric Bodies

Animal Kingdom
Internalised Physical, Etheric & Astral Bodies

Human Kingdom
Internalised Physical, Etheric, Astral & Spirit Bodies

strongly enough a plant will not flower. Plants do not have to 'reach puberty' and flower in the same way animals and humans must. We will see more examples of this later.

It is in the animal kingdom that the Astral Body imbues the physical and life form with sensation and movement. Commonly called the sense body, the Astral enables the living forms to experience their environment through taste, smell, touch and sound. From this sensation, the entity can then determine whether it feels good to be at that point or not. This is the 'animal' intelligence (instinct) carried by the Astral, which responds to stimuli - Does it feel good or not? The Astral body is also necessary for waking and sleeping to take place. Only animal and human kingdoms can really experience awakening and sleep. In this sense the plant kingdom is in a constant state of 'human' sleep. Plant processes just speed up and slow down according to the external stimuli of hot\cold, light\dark, wet\dry.

The Astral's hallmarks in the animal kingdom are the formation of true proteins, the degree of sensitivity and movement the animal possesses, the ability to internalise the breathing, along with the formation of true organs and the animals reactions to light. (3)

In the plant world, the Astral's influence shows in the degree of leaf segregation, the intensity of flowering and the development of poisonous nitrogen based alkaloids and proteins. The Solanance and Legume families are two which draw the Astral activity closely into themselves. Hence their poisonous hallucinogenic effects. The states of consciousness some substances from these families produce in Humans, gives experiences of the Astral realm.

In Humans, the Astral influences the formation of protein and the organs, however it also makes its presence felt through our psychological sphere. Emotions, dreams, imaginations and 'psychological ' swings are all related to the Astral body's activity. A careful study of the relationship between planetary movements and one's psychological changes, is a worthwhile aid in the path towards the Spirits objectivity. As it is the Astral body and its wanderings that the Spirit needs to gain objectivity with.

The Astral body resides cosmically in the spheres of the visible planets. As such, a sevenfold character can be seen as its imprint. While the physical organs are related to various planets, the seven energy centres of the body referred to as the Charkas are energetic 'organs' of the Astral body.

The Ego, or internalised spirit, imbues the sensitive life form with objective intelligence and individuality. This is separate from the dream consciousness of the astral intelligence or instinct. Ego intelligence is related to processes of thought and deduction, as well as remembering and forgetting. Processes which enable its recipient to consciously determine action and response. Once the response has been determined, the Ego imparts the degree of commitment by which the action is carried out. Through this action body heat is created. An over active Spirit in turn shows as obsessional behaviour.

The Humans are the only kingdom presently with the potential to internalise this body. This bestows upon us the potential for self-consciousness and to make free choice with regards to the sensations and instincts the Astral body experiences. The degree this free choice is exercised is determined by how 'possessed' the Spirit is, by the Astral body.

In the other kingdoms of nature, the Spirit works from outside, as a collective function called the group soul. In animals, the group soul is evident in the flock of sheep or birds, the school of fish and the pack of dogs. In both animals and plants, the species type connects the individual plant to the collective Ego/Spirit. It is fair to imagine that every species is sourced from an individual Fixed Star, while variations within the species are due to subsequent planetary and atmospheric influences, collected along its journey to Earth.

With Humans, internalising the Spirit properly, allows our connection through race and blood ties to begin to diminish in importance. We become individuals of the one human species, able to form associations through ideology, faith and individual preference, irrespective of race and blood background.

While these are the four major activities that influence life's functions, in creation there is very little black and white and many shades of grey. In observing how various living entities work with these four bodies , there are often occasions when an entity is in a grey area. Certain animals may seem very plant like, or almost human, while even humans can sometimes take on a plant quality. Nevertheless, they will be an expression of these four activities interaction.

Health in all living beings arises when the relationships of the bodies are in their right order. When a disturbance arises, we see illness.

In man and animals, the astral body is connected with the physical body through the etheric body and a certain connection is the normal state. Sometimes, however, the connection between the astral body and the physical body (or one of the physical organs) is closer than would normally be the case; so if the etheric body does not form a proper "cushion" between them, the astral intrudes itself too strongly into the physical body. It is from this that most diseases arise.

We can say that the human being stands before us in physical, etheric and astral bodies, and an ego organization. In waking life these four members of the human organization are in close connection. In sleep the physical body and etheric body are together on one side, and the ego organization and astral body on the other side. With knowledge of this fact we are then able to say that the greatest variety of irregularities can appear in the connection of ego organization and astral body with etheric body

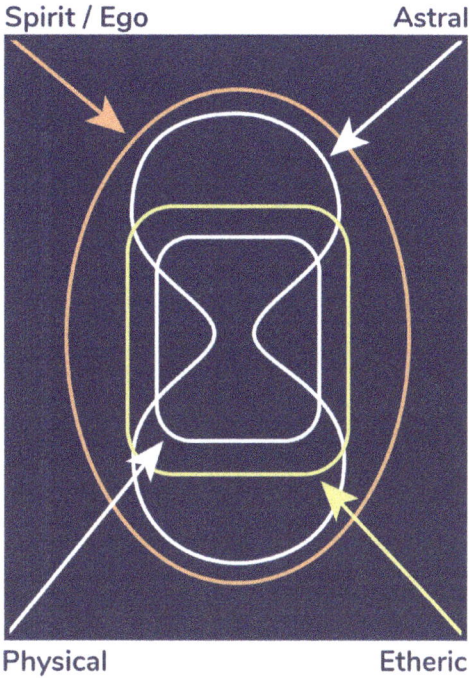

Spirit / Ego   **Astral**

Physical   **Etheric**

and physical body. For instance, we can have: physical body, etheric body, astral body, ego organization. (Plate I, 1) Then, in the waking state, the so-called normal relation prevails among these four members of the human organization.

But it can also happen that the physical body and etheric body are in some kind of normal connection and that the astral body sits within them comparatively normally, but that the ego organization is somehow not properly sitting within the astral body. Then we have an irregularity that in the first place confronts us in the waking condition. Such people are unable to come with their ego organization properly into their astral body; *thus the Ego's organising influence is very weak and the 'astral escapes its keeper'*, therefore their feeling life is very much disturbed. They can even form quite lively thoughts. For thoughts depend, in the main, upon a normal connection of the astral body with the other bodies. But whether the sense impressions will be grasped appropriately by the thoughts depends upon whether the ego organization is united with the other parts in a normal fashion. If not, the sense impressions become dim. And in the same   measure that the sense impressions fade, the thoughts become livelier. Sense impressions can appear almost ghostly, not clear as we normally have them. The soul-life of such people is flowing away; their sense impressions have something misty about them, they seem to be continually vanishing. At the same time their thoughts have a lively quality and tend to become more intense, more colored, almost as if they were sense impressions themselves.

When such people sleep, their ego organization is not properly within the astral body, so that now they have extraordinarily strong experiences, in fine detail, of the external world around them. They have experiences, with their ego and astral body both outside their physical and etheric bodies, of that part of the world in which they live — for instance, the finer details of the plants or an orchard around their house. Not what they see during the day, but the delicate flavor of the apples, and so forth. That is really what they experience. And in addition, pale thoughts that are after -effects in the astral body from their waking life. (1)

With the plant and animal the relationship of the bodies is somewhat different.

What is the human brain for? It *acts* as a support for the Ego. The animal, let it be remembered, has as yet no Ego; its brain is only on the way to Ego-formation. In man it goes on and on to the complete forming of the Ego. How then did the animal's brain come into existence? Let us look at the whole organic process. All that which eventually manifests in the brain as Earthly Matter *from the Physical body*, has simply been "excreted , (deposited), from the organic process. Earthly Matter has been excreted in order to serve as a base for the Ego. Now the process of the working-up of the food in the digestive tract and metabolic and limb system produces a certain quantity of Earthly Matter, which is able to enter into the head and to be finally deposited as Earthly Matter in the brain. But a portion of the food stuff is eliminated in the intestine before it reaches the brain. This part cannot be further transformed and is deposited in the intestine for ultimate excretion.

We come here upon a parallel which will strike you as being very paradoxical but which must not be over-looked if we wish to understand the animal and human organisations. What is brain matter? It is simply the contents of the intestines brought to the last stage or completion. Incomplete (premature) brain-excretion passes out through the intestines. The contents of the intestines, are in their processes, closely akin to the contents of the brain. One could put it somewhat grotesquely by saying that that which spreads itself out in the brain is a highly advanced dung-heap. And yet the statement is essentially correct. By a peculiar organic process, dung is transformed into the noble matter of the brain, there to become the foundation for the development of the Ego, *by providing a physical structure to receive the Cosmic Forces, the Ego uses.* In man the greatest possible quantity of intestinal dung is transformed into cerebral excrement because man bears his Ego on the earth. In animals the quantity is less. Hence there remain more forces in the intestinal excrement, of an animal which we can use for manuring. In animal manure, there is therefore more of the potential Ego element, since the animal itself does not reach Ego - hood. For this reason animal dung and human dung are completely different. Animal dung still contains Ego-potentiality. In manuring a plant, we bring this Ego-potentiality into contact with the plant's root. Let us draw the plant in its entirety (Diagram 16). Down here you have the root; up there the unfolding leaves and blossoms. And as above, in the leaves and blossoms, the astral element (red) is acquired from contact with the air, so the Ego-potentiality (orange) develops below in the root through contact with the manure. *This Ego potentiality is also recharged during the winter stage of the Silica cycle, we will look into in a later session.*

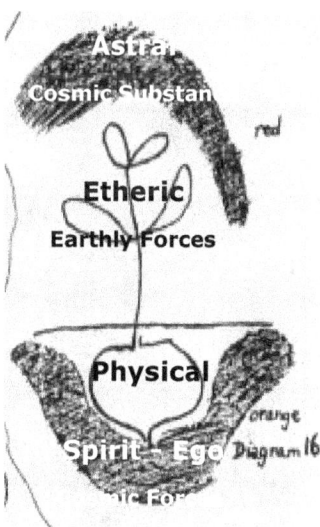

The farm is truly an organism. The astral element is developed above, and the presence of orchard and forest assists in collecting it. If animals feed in the right way on the things that grow above the earth, then they will develop the right Ego-potentiality in the manure. If they produce, this

Ego-potentiality, it will work on the plant from the root, will cause it to grow upwards from the root in the right way according to the forces of gravity. It is a wonderful interplay, but in order to understand it one must proceed step by step.

Let us recall what I said about the plant as having a physical and etheric body and being more or less surrounded from above by the astral element. The plant does not reach the astral element but is surrounded by it. If the plant enters into a special relation with the astral element, as in the case of the formation of edible fruits, a kind of food is produced which will strengthen the astral element in the animal and human organism.

QUESTION: Has liquid manure the same force of ego-organisation as dung?

ANSWER: Of course, liquid manure and dung should be used in union with each other and both should contribute to the same force of organisation of the soil. The connection with the Ego to which I referred holds good particularly for the dung, but does not hold good in general for the liquid manure.   For every Ego, even in the rudimentary form in which it appears in manure, must work in conjunction with some astral element, and the dung would have no astrality unless the liquid manure were there. The liquid is strong is astrality, the dung in ego-force. The manure may be regarded as "grey matter," while the liquid is the cerebral fluid.

Pest Control

Now I am going to tread on very thin ice and take an example very near home. I am going to talk about the nematode of the beetroot. The outer signs of this disease are a swelling of root fibers and limpness of the leaves in the morning. Now we must clearly realise the following facts: The leaves, the middle part of the plant which undergo these changes, absorb *astral* influences that come from the surrounding air, whereas the roots absorb the Ego forces which have entered into the earth and are reflected upwards into the plant. What, then, takes place when the nematode occurs? It is this: The process of *astral* absorption which should actually reside in the region of the leaves has been pressed downwards and embraces the roots.

Thus if this (Diagram No. 10) represents the earth level, and this the plant, then in the plant infested with the nematode ,the *astral cosmic substance* forces which should be active above the horizontal line are actually at work below it. What happens is that certain *astral* forces slide down to a deeper level; hence the change in the external appearance of the plant. But this also makes it possible for the parasite to obtain under the soil (which is its proper habitat) those *astral* forces which it must have to sustain it (the nematode is a wire-like worm).

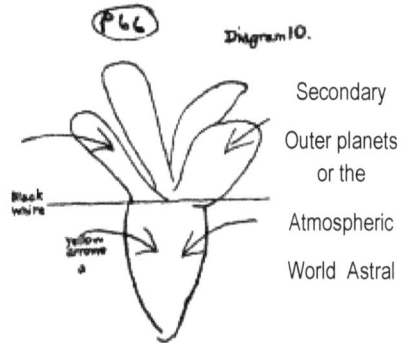

Diagram 10.

Secondary

Outer planets or the

Atmospheric

World Astral

Otherwise it would be forced to seek for these forces in the region of the leaves; this, however, it cannot do as the soil is its proper environment. Some, indeed all, living beings can only live within certain limits of existence. Just try to live in an atmosphere 70 degrees above or 70 degrees below zero and you will see what will happen. You are constituted to live in a certain temperature, neither above nor below it. The nematode is in the same position. It cannot live without earth and without the presence of certain *astral* forces brought down into it. Without these two conditions it would die out.

Our task, is to help the energetic activities back to their rightful place, and the pest will have no choice but to move away.

Every living being is subject to quite definite conditions. And for the particular beings with which we are dealing, it is important that *astral* forces should enter the earth, forces which would ordinarily display themselves only in the atmosphere around the earth. Actually the workings of these forces have a four-year rhythm. Now in the case of the nematode, we have something very abnormal. If one enquires into these forces, one finds that they are the same as those at work on the cockchafer grubs; and as those, too, which bestow on the earth the faculty of bringing the seed potato to development. Cockchafer grubs as well as seed potatoes are bred by the same forces, and these forces recur every four years. This four yearly cycle is what must be taken into account not with regard to the nematode but with regard to the steps we take to combat it."

A particularly ticklish question was raised in the discussion we had the other day as to whether parasites could be combated in this way. i.e. by methods of mental concentration and the like. There is no doubt that if one sets about it in the right way one can do such things. The period lying between the middle of January and the middle of February is that in which the forces which have been concentrated inside the earth are most powerfully unfolded. If we were to set this period aside as it were as a festival season and undertook these acts of concentration, then we should be able to bring about such effects. When we meditate we enter into a new relationship with the nitrogen, the substance which contains the "Imaginations", *due to it being the physical carrier of the astrality.* We enter upon a state in which such things can become operative; upon a state in which we confront quite differently the whole world of plant-growth. Such effects are not so obvious today as they were in the past when these things were recognised. For there were times when people knew that by a certain inner attitude they actually fitted them-selves for the care of the growth of plants. Nowadays these delicate and subtle influences are overlooked, the presence of other people disturbs them, as is bound to happen when one is constantly moving about among people who disregard such things, this is why it is so easy to refute their existence. I therefore hesitate to talk freely of such things before a large audience, because they can so easily be refuted on the basis of the present conditions of daily life. As I said, it is a ticklish question, but a question which does admit of a positive answer. But this activity must be undertaken in harmony with the whole of Nature. One must realise that it makes all the difference whether an exercise of concentration is carried out in mid-winter or in mid-summer.

Plants

I shall ask you to-day to join me in the consideration of rather more recondite matters, to follow me into what is nowadays an almost unknown territory, although the instinctive husbandry of the past was thoroughly conversant with it. The beings in Nature - minerals, plants, animals - we will disregard man for the moment - are often regarded as though they existed in completely separate realms. It is the custom to-day to look at a plant as though it existed by and for itself, and similarly one species of plant is also regarded as being isolated from other plant species. So these things are neatly sorted and fitted into general species, as though they were being put into boxes. But things are not like this in Nature. In Nature - nay, in the world-being as a whole, all things are in mutual interaction. One thing is always being affected by another. In these materialistic days, only the more palpable effects of this interaction are noted, such as when one thing is eaten or digested by another, or when the dung of animals is used for the soil. In addition to these, however, finer interactions amongst more delicate forces and substances are continually taking place: through warmth, through the chemical-etheric element which is continually at work in the atmosphere, and through the life-ether. Unless we take account of these more delicate interactions, we shall make no progress, at any rate in certain departments of Agriculture. In particular we must look to those more intimate interactions which take place in Nature when we have to deal with the life together of plant and animal on the farm. We must look with understanding not only upon those animals which undoubtedly stand close to us, such as cattle, horses, sheep, etc., but also, for example, upon the manifold insect world which during a certain period of the year hovers around the plants. Indeed we must learn to look with understanding at bird life too. Humanity to-day is very far from realising how much farming and forestry are affected by the expulsion from certain districts, of certain kinds of birds as a result of modern conditions. Here again light can be thrown on the subject by conclusions given by Spiritual Science. Let us therefore extend some of these ideas which have been working upon us and come by their help to a yet wider vision.

A fruit tree - apple, pear or plum - is something completely different in kind from a herbaceous or cereal plant as any kind of tree outwardly is indeed. But, putting aside any preconceived notions, we must find out wherein the peculiarity of the tree lies. Otherwise we shall never understand the function fulfilled by fruits in the economy of Nature. I am speaking, of course, of the fruit that grows on trees. If we look at a tree with understanding we shall find that the only parts of it which can really be reckoned as plant are the tender twigs, the green leaves and their stalks, the blossoms, the fruits. These grow out of the tree just as herbaceous plants grow out of the soil, the tree being in fact "earth" in relation to the parts that grow out of it. It is as though the soil were heaped up - but a somewhat more quickened soil than the ordinary soil in which our herbaceous and cereal plants grow.

If, therefore, we want to understand the nature of a tree, we must observe that it consists of the thick trunk, to which are attached the branches and boughs. On this ground the specifically plant-like parts grow, viz. leaves and blossoms, which are as much rooted in the trunk and branches as cereal and herbaceous plants are rooted in the earth. The question therefore arises: Is this plant– or plant like part - which may be regarded as more or less parasitical, really rooted in the tree?

We cannot discover an actual root on the trees. We conclude, therefore, that this plant, which develops its leaves and blossoms and twigs up aloft, must have lost its roots in growing on the tree. But no plant is complete without its root. It must have a root. Where, then does the actual root of this plant reside?

Now, the root is only invisible for our limited outer vision. In this case one does not see it, but has to understand where it is. What do we mean by this? The following concrete comparison may help. Imagine a large number of herbaceous plants so closely together that their roots were intertwined and grew into each other, forming a completely matted mass or pap of roots. You can well imagine that this pap does not remain chaotic, but that it organises itself into a unity so that the sap-bearing vessels unite with each other. In this organised root-pap, it would not be possible to distinguish where one root finished and the other began, and a common root-organ would arise (See Diag. No. 12). A thing like that does not, of course, exist in the soil, but such a root-formation is actually present in the tree. The plants that grow on the tree have lost their root, have become relatively separated from it and are only, as it were, etherically connected with it. What I have drawn hypothetically is really the layer of cambium (a layer of living cells lying between the last-formed wood and the outer bark) in the tree and we cannot regard the roots of these plants otherwise than as having been replaced by the cambium. From this tissue, which is always forming new cells, these plants unfold themselves just as from the root below a herbaceous plant unfolds above the soil. We can now begin to understand what the tree really is. The tree with its cambium - which is the only soil-producing layer in the tree, is actually heaped-up earth, which has grown upwards into the air element and therefore requires a more interiorised form of life than is present in the ordinary soil which contains the root. Thus we must regard the tree as a very curious entity, whose function it is to separate the "plants" growing on it (twigs, blossoms, fruit; from their roots; an entity which places between them and their roots a distance which is bridged only by spirit - or more strictly by the Etheric. It is in this way we need to look, with a macro-cosmic understanding, into the facts of growth.

But the matter goes much farther. What results arise from the existence of a tree? That which is around the tree in the air and outer warmth is of a different plant-nature from that which grows up from the soil in the air and warmth and forms the herbaceous plant. It is a plant-world of a different order, possessing a far more intimate relation with the surrounding astral element. Lower down that element is eliminated from the air and warmth in order to make them mineral-like, so that they can be used by man and beast (see lecture II. They become "dead" air and warmth).

It is true, as I have said, that the plant we see growing upon the ground is surrounded, as with a cloud, by the astral element. But around the tree, the astral element is far denser. So much so, that we may say: Our trees are definitely collectors of astral substance.

Here one might say it is quite easy to reach a higher development and, become "esoteric" - I do not mean clairvoyant but clairsentient as to the sense of smell. One has only to acquire the capacity for distinguishing between the scent of plants growing in the ground, the peculiar smell of orchards, especially in the spring when they are in flower, and the aroma of forests. *Then one is able to tell the difference between* a plant atmosphere poor in astral elements, such as that of herbaceous plants growing in the soil and an atmosphere such as we sniff with such pleasure when the scent of trees is wafted in our direction. And if you train your sense of smell to distinguish between the scent of soil-grown (herbaceous) plants and the scent of trees, you will have developed "clear-smelling" for the thinner and for the denser forms of the astral element. The countryman, as you see, can very easily acquire this "clear-smelling" though this faculty, common in the old days of instinctive clairvoyance, has been much neglected in recent times.

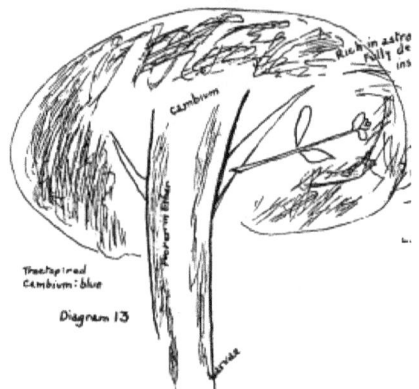

Diagram 13

If, now, we realise the consequences to which this may lead the question will arise: what is happening in that part of the tree which may be regarded as the opposite pole from the "parasitical" plants on the tree which collect this astral element? What is happening through the cambium?

Now, the tree makes the atmosphere far and wide around it richer in astral element. What happens while the "parasite" growth goes on above in the tree? The tree here has a certain inner vitality, a powerful etheric life in it. The cambium tones down this vitality, making it more mineral in nature.

While about the upper part of the tree an enrichment of the astral substance is going on, the cambium causes an impoverishment of the etheric life in the tree. The tree within is deprived of etheric life as compared with the herbaceous plant. In consequence, this produces a change in the root. The root of the tree becomes more mineral, far more mineral than the roots of the herbaceous plants. But by becoming more mineral, the tree root withdraws some of the etheric life from the soil; it makes the soil around the tree slightly more dead than it would be around a herbaceous plant. This must be fully borne in mind, for these natural processes always have a great significance in the economy of Nature. We must therefore seek to understand the significance of the astral wealth in the atmosphere around the tree and of the etheric poverty in the region of the roots.

If we look around us, we can find the further connection. It is the fully developed insect which lives on and weaves in this enriched astral element which wafts through the trees; whereas the impoverished etheric element beneath, spreading in the soil and throughout the whole creation, is that which harbours the larvae or grub. Thus if there were no trees on the earth there would be no insects. The insects that flutter around the upper parts of the trees and through the forests depend for their life upon the presence of the trees; and exactly the same thing is true of the grubs.

Here we have yet another indication of the inner connection between all roots and animal life beneath the soil. This is especially evident in the case of the trees. But this same principle which is so striking in the case of the trees is present in a modified form throughout the whole of the vegetable world, for in every plant there lives something that tends to become a tree. In every plant the root and what is around it tends to throw off the etheric life whereas the upper growth strives to attract the astral element more closely to itself. For this reason there arises in every plant that kinship with the insect world which I have specially characterised in the case of the tree.

Worms

This relation, however, to the insect world in fact extends so as to comprise the whole of the animal world. In former times insect grubs, which can only live upon the earth because of the presence of tree roots, transformed themselves into other kinds of animals, similar to larvae and remaining at the larva stage throughout their lives. These animals then emancipated themselves to a certain extent from the tree root nature and adopted a life which extends also to the root region of herbaceous plants. And now we find the curious fact that certain of these sub-terrestrial animals, though far removed from being larvae yet have the ability to regulate the amount of etheric life in the soil if this amount becomes excessive. When the soil becomes, as it were, too much alive and the sprouting etheric life too strong, these animals of the soil see to it that this excess is reduced. They are thus wonderful vents which regulate the vitality in the soil.

These lovely creatures, for they are of the greatest value to the earth, are no other than the common earth-worms. One ought to study the life of earthworms in relation to the soil, for these wonderful animals allow just that amount of etheric life to remain in the soil as is needed for the growth of plants. Thus in the soil we have these creatures, earth-worms and their like, distantly resembling larvae. One ought in fact to see to it that certain soils which require it are supplied with a healthful stock of worms. We should soon see how beneficent such a control over this animal-world in the soil can be, not only for vegetation but also thereby for the rest of the animal kingdom, as we shall show later.

Birds

Now there are certain animals which bear a distant resemblance to the insect world, to that part of it which is fully developed and winged, I mean the birds. It is well known that in the course of the development of the earth something very wonderful took place between the birds and the insects. It is as though, to put it figuratively, the insects had one day said: "We do not feel strong enough to "work-up" the astrality sparkling around the trees, we shall therefore use the "desire to be a tree" of other plants. We shall flutter around these, and leave largely to you birds the astral life that surrounds the trees." Thus there arose in Nature a proper "division of labour" between the birds and the butterflies; and this co-operation in the winged world brought about in a wonderful manner the right distribution of astral life wherever it was required on the surface of the earth. If these winged creatures are removed, the astral life will fail to accomplish its proper function, and this will be noticeable in the stunted condition of the vegetation. The two things are connected; the world of winged animals and all that grows out of the soil into the air. The one is unthinkable without the other. In farming, therefore, we must see to it that birds and insects fly about as they were meant to do and the farmer should know something about the breeding and rearing of birds and insects. For in Nature - I must repeat this again and again - everything, everything is connected.

These considerations are of the utmost importance for a right understanding of the questions before us and we must therefore hold them very clearly in our minds. The winged world of insects brings about the proper distribution of astrality in the air. The astrality in the air has a mutual relationship with the forest which directs it in the proper way, much as in the human body the blood is directed by certain forces. And this activity of the forest, which is effective over a very wide area, will have to be undertaken by something quite different in a district where there is no forest. Indeed, in districts where woods alternate with arable land and meadows that which grows in the soil comes under quite different laws from those which rule in completely unwooded districts.

There are certain parts of the earth which were obviously wooded areas long before man took a hand. In certain matters, nature is cleverer than we are, and it may safely be assumed that if a forest grows naturally in a certain district it will have its uses for the neighbouring fields and for the herbaceous and cereal vegetation round about. In such districts one ought therefore to have the intelligence not to uproot the woods but to cultivate them. And as the earth is gradually changing through climatic and cosmic influences of all kinds, one should have the courage, when the vegetation becomes poor, not merely to indulge in all sorts of experiments in the fields and for the fields, but to increase the area of woods in the neighbourhood. And when plants run to leaf, lacking the power to produce seed, one should take bites out of the neighbouring woods. The regulation of woods in districts which Nature intended to be wooded is an integral part of agriculture, and must be examined with all its consequences from a spiritual point of view.

Again, the world of grubs and worms may be said to stand in a mutual relationship to the lime, i.e. to the mineral part of the earth; while the world of birds and insects, of all that flies and flutters about, has a similar relationship to the astral element. The relationship between the worm and grub world and lime brings about the drawing off of the etheric element. This is the function of lime, but it performs this function in co-operation with the world of worms and grubs.

If things are to be rightly handled, it is necessary to gain insight into the mode of activity of substances *(physical)* , and forces *(etheric)*, the dynamic *(astral)* and of the spiritual too in every part of agriculture. A child who does not know what a comb is for, will bite into it or otherwise misuse it. In the same way we shall make quite a wrong use of things if we do not understand their essential being and their specific functions.

Compost and Manure

To make the matter clearer let us take the case of a tree *again*. A tree is different from an ordinary annual plant which remains at the merely herbaceous stage. It surrounds itself with rind and bark, etc What then is the fundamental nature of the tree as opposed to that of an annual plant? In order to answer this question, let us compare the tree to a mound of soil which has been piled up and is exceptionally rich in humus, i.e. which contains an exceptionally large quantity of more or less decomposed vegetable matter, and includes perhaps some decomposing animal matter as well. Let us assume that this is the mound of soil, rich in humus, and I will make in it a

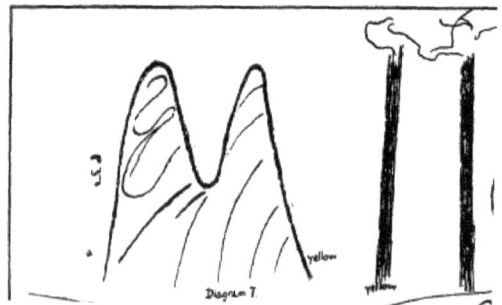

Diagram 7.

crater-like depression; and let us take this (Indicated in the second part of the drawing) as the tree, the more or less solid part being outside, while inside grows that which goes to build up the tree as a whole. It may strike you as strange that I should place these two things side by side, but they are more closely related than you may perhaps think. The reason is that soil such as I have described, soil containing plenty of humus, i.e. substances in course of decomposition, bears etheric life within it. And this is the point. When soil is so constituted as to have etheric life within it, it is on its way to becoming the outside covering of the plant, but does not in fact develop so far as to become bark. Now imagine (although, of course, this does not happen in Nature) that such a mound of soil, with its humus content has, by means of its etheric life, raised itself to a higher form of development and wrapped itself round the plant. For if any part of the earth is raised above the general level, if the outer separates itself from the inner, then that which is raised above the normal level will show a definite tendency to life, a distinct tendency to be penetrated with etheric life. This is why, if you want to make inorganic soil more fertile by mixing it with humus-like substance or with any sort of decomposing refuse, you will find it easier to do so successfully if the soil is heaped up into mounds. For then the soil itself will have the tendency to become inwardly alive and plant-like. The same process takes place in the formation of a tree. The soil bulges upwards, as it were, and surrounds the plant with its own etheric life. Why do I say this? The reason is that I wish to waken your consciousness to the fact that there is an intimate kinship between what is enclosed within the contours of the plant and that which comprises the soil round the plant. It is untrue that the life of the plant stops short at its outer sphere. The actual life is continued, particularly from the roots, into the soil and in many cases there is no sharp boundary between the life within the plant and that in its immediate environment.

In order to have a fundamental understanding of a soil which is manured or similarly treated, one must know that manuring consists in a vivifying of the soil so that the plant may not be planted in dead soil. A plant will more easily develop from its own vitality, for what is necessary for fruit formation, if it is planted in something already alive. Fundamentally all plant growth is slightly parasitic in character; it grows like a parasite on the living earth. And it must be so. In many parts of the earth we cannot rely on Nature herself to supply a sufficient quantity of waste organic matter to enable the soil adequately to revivify itself by decomposition of such matter. In those places, therefore, we must assist the growth of plants with manure. This necessity, however, arises least of all in districts containing so-called "black soil", for here Nature herself has seen to it that the soil is sufficiently alive.

You will see from all this what is really happening; but there is something further which must be understood. One must learn - and this may not always be pleasant – to enter into a personal relationship with everything that comes within the sphere of Agriculture, and particularly with the work connected with manure and manuring. The job may seem to be an unpleasant one, but you cannot do without this personal relationship. Why? Well, if you consider the nature of any living being, you will find the reason. Every living being always has an inner and an outer side. The inner side is inside some kind of skin, the outer side is outside that skin. Let us begin with the inner side.

The inner side of every living thing has not only streams of force which go outwards in the direction shown by these lines but it also has streams of force which go inwards from the skin, which are pressed back. Now an organism is surrounded on the outside by streams of all kinds of forces. There is something which expresses very exactly although in a "personal" way the relationship which must be established by the organism between its inner and outer side. All the forces working inside the skin, all that stimulates and maintains life, must - pardon the phrase - inwardly smell, must have an inward stench. Taken as a whole, life itself consists in this that what is generally diffused as a scent is instead held together so that the scent is kept inside and does not stream outwards too strongly. An organism must therefore allow as little as possible of its scent-producing life to escape outwards through its skin. Indeed one

might say that the healthier an organism, the more it will smell inwardly and the less it will smell outwardly. A living organism and particularly the plant organism (apart from the flower) is designed not to give out scent but to take it in. And if we consider the beneficial influences on a meadow full of fragrant aromatic flowers, we shall begin to notice how living things mutually support one another in Nature. This fragrance of flowers which is diffused and which is something different from the odour of mere life, issues from sources of which we shall become aware later and it acts on the plants from outside. One must enter into a personal, living relation to all these things; only then are we really one with Nature.

Now the main thing to understand is that manuring and the like must consist not only in conveying a certain degree of aliveness to the soil, but also in enabling the nitrogen to spread through it, in such a way that with its help the life is carried along certain lines of force as I showed yesterday. In manuring therefore we must bring sufficient nitrogen into the soil to enable the life to be borne into the organic structure of the soil which is to bear the plant. This is the task, but it must be carried out exactly and properly.

Now here is a very significant hint: when purely mineral matter is used for manure, it never reaches the earth element, but at best only the water element in the soil. You can produce with mineral manures an effect in the watery part of the earth, but you will not achieve a vivification of the earth element itself. Plants therefore, which are under the influence of any sort of mineral manure will exhibit a type of growth which betrays that it comes from water which has been activated, not from the solid element which has been vivified. The best way to approach these things will be to take the most unassuming and often despised kind of manure, viz. compost. Here we have a means of vivifying the soil. We include in compost all kinds of neglected refuse from farm or garden, mown grass, fallen leaves, and the like, nay, even to the remains of dead beasts, etc. These things should by no means be despised, for they retain something not only of the etheric but even of the astral elements. And that is important. In a compost heap, all contained in it is actually pervaded not only by living and etheric but also by astral elements. These are present to a lesser degree in solid or liquid animal manure, but they are more stable, more settled - especially the astral element only we must make use of this stable or settled character in the right way. The action of the astral element upon nitrogen is hindered wherever the etheric element is too ebullient.

A too powerful sprouting of the etheric life hampers the astral element in the compost heap from doing its work. Now there is in Nature a substance which I have already mentioned from varied angles which is extremely useful in this respect, and that is the chalky or limestone element. If therefore, some of this - preferably in the form of quicklime - is introduced into the compost heap, we get the following special result: without causing the astral element to "volatilise" as it were too much, the etheric element is taken up by the quick-lime and the oxygen is absorbed as well; In this way, the astral element is brought to a wonderful activity. This leads to a very definite result: In manuring the soil with compost, we are giving over to it something which has the tendency to carry the astral element directly into the solid element without the detour through the etheric element. In this way, therefore, the earthly element is thoroughly "astralised" and thereby becomes penetrated with nitrogen. This result, indeed, very much resembles a certain process in the human organism - a plant-like process - so plant like in fact that it does not proceed to fruit formation, but stops at the stage of leaf and stem formation. What we give over to the soil in the compost has its parallel in that process which brings about in the food we eat that "mobility" of which I spoke before. We bring about a similar activity in the soil when we treat it in the manner described. Soil prepared in this way will be especially suitable for producing plants which, when they are eaten by animals, will continue to bring about a similar activity in their organisms. In other words, we shall do well to manure our meadows and pasture lands with this compost, and if we carry through the process carefully, with strict regard for the other proceedings and ingredients, we shall succeed in obtaining good fodder, which, when mown and dried, preserves its quality. I should like to remind you that to take the right steps, one must look into the nature of the whole process, and finding the right thing to do in any particular case will, of course, depend to a great extent upon having the right feeling. This feeling, however, develops, when we look into the whole nature of this compost process. For instance, if the compost heap is left alone the astral element in it will begin to spread in all directions. It will then be a question of developing the right personal relation to the heap in order to find out how it can be made to retain its smell within it. This can easily be done by putting down a thin layer of the compost material and covering it with peat moss, then adding another layer and so on.

In this way we hold together that which would otherwise "volatilise" itself as smell. Nitrogen, indeed, is a substance which in all its modifications is eager to spread out into all directions. And now it is held back, by this I wish to indicate how necessary it is to treat the whole "agricultural-individuality" in the light of the conviction that etheric life and even the astral principle must everywhere be poured out over it to make our work effective.

The animal organism is *also* connected with the whole economy of nature. With respect to form and colour structure and consistency of its substance, it is under the influence of the planets. Working backwards from the snout, the influences are as follows , Saturn, Jupiter and Mars affect the region extending from the snout to the heart, the heart is worked upon by the Sun, while the region extending from behind the heart to tail comes under the influences of Venus, Mercury and Moon.

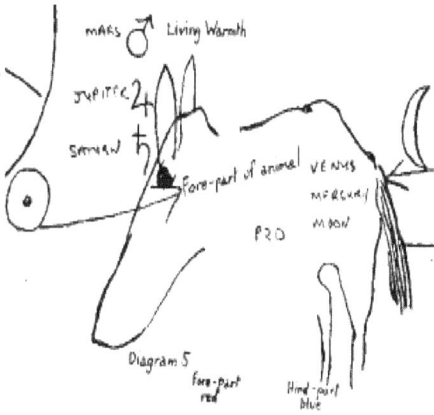

Diagram 5

Those who are interested in these things should try to examine the forms of animals from this point of view. For a development of knowledge along these lines would be of enormous importance. Go to a museum, for example, and examine the skeleton of any mammal. In doing so, bear in mind the principle that the structure and build of the head is primarily the result of the direct radiation of the Sun streaming into the mouth.

Then you will see that the structure of the head and of the adjoining parts depends upon the way in which the animal exposes itself to the Sun. A lion exposes itself quite differently from a horse: the reason for these differences will be examined later on. Thus the front part of an animal and the structure of its head are directly connected with the Sun's radiation. Now the light of the Sun also reaches the Earth indirectly, by being reflected from the Moon. This too has to be taken into account. The sunlight that is reflected from the Moon is quite ineffectual when it falls on the head of an animal. (These things apply especially to embryonic life). The light reflected from the Moon produces its greatest effect when falling upon the hind parts of the animal. Look at the formation of the skeleton of an animals hind parts and the peculiar polarity in which it stands to the formation of the head. You should develop a feeling for this contrast in form between the animals hind quarters and its head and especially for the insertion of the hind limbs and the rear and the intestinal tract. This contrast between the front and the hindermost parts of the animal is the contrast between Sun and Moon. If you go further you will find that the influence of the Sun stops just short of the heart; that Mars, Jupiter and Saturn are active in the formation of the blood and the head; and that, from the heart backwards the activity of the Moon is reinforced by that of Mercury and Venus.

In a similar manner we can look at plants and the soil. If then, certain forces coming from the Moon, Venus and Mercury enter the Earth and become effective in plant life, the question arises: What will promote and what will restrain the activity of these forces?

This soil - I will indicate it schematically by this straight line (see diagram no. 2) is generally looked upon as being something purely mineral into which at the best organic substance has entered either because humus has been formed or manure has been introduced. The idea that the soil not only contains added organic substance but also has itself a plant - like nature - and even contains an astral activity; such an idea has never been considered, still less conceded.

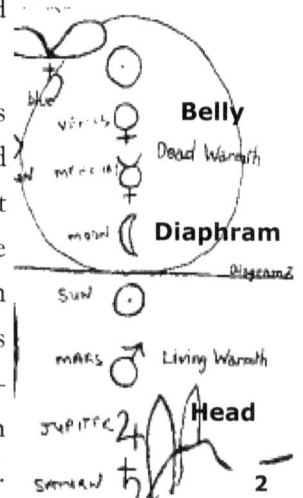

82

And if we go a step further and consider how this inner life of the soil in the delicate balancing of its distribution is quite different in summer from what it is in winter, we come to subjects which are of enormous importance in practical life to which no attention is paid to-day. If you start by considering the soil then you must bear in mind the fact that it is a kind of organ within that organism which manifests itself wherever the growth of nature appears. The earth surface is really an organ, an organ which, if you care to, you may compare with the human diaphragm. We may put the matter broadly in this way (it is not quite exact but will give the right idea): Above the diaphragm there are in man certain organs, the head in particular, and the processes of breathing and circulation which work up into the head. Under the diaphragm are other organs. Now if we compare the earth surface with the human diaphragm we must say: The individuality represented by our farm, having the earth surface for its diaphragm has its head under the earth, while we and all the animals live in its belly. Above the surface of the earth, is really what may be regarded as the bowels of what I will now call "agricultural-individuality". On a farm we are walking about inside the belly of the farm, and the plants grow upwards within this belly. Thus we are dealing with an individuality which is standing on its head, and which is only rightly looked at if so understood, especially as regards its relation to Man. In relation to animals, the situation, as we shall see later on, is slightly different.

Cowhorns

Now following the trend, *we saw in compost,* we can take a further step. Have you ever wondered why it is that cows have horns, while certain other animals have antlers? It is a very important question. Yet what science has to say about it is quite one-sided and based on externals. Let us consider why cows have horns. I said that the forces within a living organism need not always be directed outwards, but can also be directed inwards. Now imagine an organic entity possessing these two sets of forces, but which is unformed and lumpish in build. The result would be an irregular, ungainly being. We should have curious looking cows if this were the case. They would all be lumpish and unformed, with rudimentary limbs as at an early embryonic stage. But this is not how a cow is constructed. A cow has horns and hoofs. Now what happens at the points where horns and hoofs grow? At these points an area is formed from which the organic formative forces, *moving outwards from the metabolism,* are reflected inwards in a particularly powerful way. There is no communication with the outside as in the case of the skin or hair; the horny substance *of the horn* blocks the way for these forces to the outside. This is why the growth of horns and claws has such a bearing upon the whole form of the animal.

Things are quite different in the case of antlers. Here the streams of forces, *coming from the metabolism* are not led back into the organism, but certain of them are guided for a short distance out of the organism; there must be valves, as it were, through which the streams localised in the antlers (we can speak of streams of 'force', just as we can speak of streams of air or liquid) can be discharged. A stag is beautiful because it stands in intense communication with its environment by reason of its sending outwards streams of *metabolic* forces; by this it lives within its environment and takes up from it everything which works organically in its nerves and senses. Hence the nervous nature of the stag. In a certain respect all animals which have antlers are suffused with a gentle nervousness. This is clearly to be seen in their eyes.

The cow has horns, in order to reflect inwards the astral and etheric formative forces, *coming first from the metabolism*, which then penetrate right *back* into the metabolic system, so that increased activity in the digestive organism arises by reason of this radiation from horns and hoofs. If one wants to understand foot-and-Mouth disease, i.e. the retroaction from the periphery to the digestive tract, one must know of this connection. Our remedy for Foot-and-Mouth disease is based on the recognition of this. In the horn, therefore, we have something which by its inherent nature is fitted to reflect the living etheric and astral streams into the inner life organs. The horn is something which radiates etheric life and even the astral element. Indeed, if you were able to enter into the cows belly, you would smell the current of etheric-astral life which streams *back* from the horns: and the same thing is true of the hoofs.

Now this gives us a hint as to the measures we may recommend for increasing the effectiveness of ordinary stable manure. What is ordinary stable manure really? It is foodstuff which the animal has taken in and which up to a certain point has been assimilated by its organism, thereby stirring into activity certain dynamic forces in the organism. Its main use has not been to increase the amount of substance in the organism, for after having had its effect, it is excreted. It has become permeated with astral and etheric elements. The astral element has filled it with nitrogen-bearing forces and the etheric element with oxygen-bearing forces. The substance which emerges as dung is permeated with these forces. Imagine now: We take this substance and pass it into the soil in some form or other (the details will be dealt with later). Thus we add to the soil an etheric-astral element whose proper place is in the belly of the animal, where it produces forces of a plant-like nature. For the forces which we produce in our digestive tract are of a plant-like nature. We should be extremely thankful that we get such a residue as dung, for it carries etheric and astral forces from the interior of the organism out into the open. These forces remain with it, and it is for us to keep them there. In this way the dung will act in a life-giving and also astralising way on the soil, not only on the water element in it, but especially on the solid element. It has the power to overcome what is inorganic in the earthly element. Now what is passed over to the soil will necessarily, of course, lose the form it originally had when taken in as food, for it has to go through an inner organic process in the metabolic system. There it enters upon a phase of decomposition and dissolution. But it is at its best just at the point where it begins to dissolve through the workings of its own astral and etheric elements. It is then that the parasites, the micro-organisms make their appearance. They find a good feeding-ground in which to develop. This is why the theory arose that these parasites are themselves responsible for the virtues in the manure. But they are only indications of the condition of the manure. If we think that by inoculating the manure with these bacteria we shall radically improve its quality, we are making a complete mistake. Externally there may seem at first to be an improvement, but in reality there is none.

World Forces into Protein

GA in Italics

The earthly and cosmic forces work in the processes of agriculture through the substances of the Earth. And we shall only be able to pass on to the difficult practical applications during the next few days if we occupy ourselves rather more closely with the question of how these forces work through the Earth's substances. But first we must make a digression and enquire into the activity of Nature in general.

You know that in terms of contemporary chemistry, the main ingredients of albumen are the four main natural substances, carbon, oxygen nitrogen and hydrogen, and , in addition, sulphur, as, so to speak, a omnipresent mediator, and homeopathic agent in the operations of the other four. *In the great spheres of nature we can identify, Hydrogen as the dominant chemical element of the cosmic spaces, populated by the stars. Nitrogen is found concentrated in the atmospheres of some planets, with our own atmosphere comprising 80% nitrogen. Oxygen, we find only in our atmosphere, at 20%, as a expression of the very life forms it helps to support, while our Earthly forms are primarily Carbon based.*

What interests us here is the fact that the function performed in the external world by C,H,O,N and their mediator sulphur is , *the same activity as* is being individualized in man through the four organic systems. You will see then that the *Spirit inspired* Ego organisation is connected with the Hydrogen in the same way that the physical organisation is connected with Carbon, the etheric organisation with oxygen and the Astral organisation with Nitrogen. The composition of the external atmosphere is of such a nature as to furnish the ratio for the connection between the astral and etheric bodies and concurrently between their partners the physical body and ego. (2)

One of the most important questions that can be raised in discussing production in the sphere of Agriculture is that concerning the significance and influence of nitrogen. But this question concerning the fundamental nature of the action of nitrogen is at present in a state of the greatest confusion. When one observes nitrogen today in the ordinary way one is only looking at the last offshoots as it were, of its activities, its most superficial manifestations. We overlook the natural interconnections within which nitrogen is at work: nor indeed can we help so doing if we remain in enclosed within one section of Nature. To gain a proper insight into these connections we must bring within our survey the whole realm of Nature and concern ourselves with the activity of nitrogen in the Universe. Indeed - and this will emerge clearly from my exposition - while nitrogen as such does not play the primary part in plant-life it is nevertheless supremely necessary for us to know what this part is, if we wish to understand plant-life.

In its activities in Nature nitrogen has, one might say, four sister-substances which we must learn to know if we wish to understand the functions and significance of nitrogen in the so-called economy of Nature. These four sister substances are the four substances which in albumen (protein), both animal and vegetable, combine with nitrogen in a way which is still a mystery for   present-day science. The four sister-substances are carbon, oxygen, hydrogen and sulphur. If we wish to understand the full significance of albumen, it is not enough to mention the ingredients hydrogen, oxygen, nitrogen, carbon; we must also bring in sulphur, that substance the activities of which are of profound importance for albumen. For it is sulphur which acts within the albumen as the mediator between the spiritual formative element, *carried by Hydrogen and* diffused throughout the Universe and the physical element, *manifest in Carbon*. Indeed, if we want to follow the path taken by the spirit in the material world, we shall have to look for the activity of sulphur. Even if this activity is not so visible as those of other substances it is still of the utmost importance because spirit works its way into physical nature *with the help* of sulphur; sulphur is actually the *facilitator* of spirit. The ancient name "sulphur" is connected with the word "phosphor" (which means bearer of the light) because in the old days man saw spirit spreading out through space in the outstreaming light of the Sun. Hence they called the substances which are linked up with the working of light into matter like sulphur and phosphorus the "light bearers". *Indeed all the 'brother' elements of Sulphur, on the third ring of the Periodic Table, Silica, Aluminium, Magnesium Sodium and Chlorine have 'light bearing ' tasks.* And once we have realised how fine is the activity of sulphur in the economy of nature we shall more easily understand its fundamental *mediating* nature, when we consider the four sister substances - carbon, hydrogen, nitrogen and oxygen and the part they play in the workings of the Universe. The modern chemist knows very little about these substances. He knows what they look like in a laboratory, but is ignorant of their inner significance for cosmic activities as a whole. The knowledge which modern chemistry has of these substances is not much greater than the knowledge we might have of a man whose external appearance we had noticed as he passed us in the street, and of whom we had perhaps taken a snapshot, whom we call to mind with the help of the snap-shot. For what science does with these substances is little more than to take snapshots of them, and the books and lectures of to-day about them contain little more than this. We must learn to know the deeper essence of these substances.

Let us therefore start with carbon, The bearing which these things have upon plants will soon be made clear. Carbon, like so many beings in modern times has fallen from a very aristocratic position to one that is extremely plebeian. All that people see in carbon now days is something with which to heat their ovens (coal) or something with which to write, graphite, Its aristocratic nature still survives in one of its modifications, the diamond. But it is hardly of very great value to us today, in this form, because we cannot buy it. Thus what we know of carbon is very little in comparison with the enormous importance which this substance possesses in the Universe. And yet, until a relatively recent date a few hundred years ago, this black-fellow - let us call him so - was regarded as worthy to bear the noble name of "Philosophers Stone".

A great deal of nonsense has been spoken about what was really meant by this name. For when the old Alchemists and their kind spoke of the Philosopher's Stone they meant carbon in whatever form it occurs. And they only kept their name secret because if they had not done so, all and sundry would have found themselves in possession of the Philosopher's Stone. For it was simply carbon. But why should it have been carbon?

A view held in former days will supply us with the answer, which we must come to know again. If we disregard the crumbled form to which certain processes in nature have reduced carbon (as in coal and graphite) and grasp it in its vital activity in the course of serving the bodies of men and animals and as it builds up the body of the plant from its own inherent possibilities, the amorphous and formless substance which we generally think of as carbon will appear as the final outcome, the mere corpse of what carbon really is in the economy of Nature. Carbon is really the carrier of *ALL the* formative processes in Nature. It is the great sculptor of form, whether we are dealing with the plant whose form persists for a certain time or with the ever-changing form of the animal organism. It bears within it not only its black substantiality, but in full activity and inner mobility it carries within it the formative cosmic prototypes, *of the Stars*, the great world-imaginations from which living form in nature must proceed. A hidden *Hydrogen* sculptor is at work *with* carbon in building up the most diverse forms in Nature, this hidden *Hydrogen* sculptor makes use of sulphur. If, therefore, we regard the activities of carbon in Nature in the right way, we shall see that the cosmic spirit, *carried by Hydrogen* is active as a sculptor "moistening" itself as it were, with sulphur and with the help of carbon builds up the relatively permanent plant form, and also the human form which is dissolved at the moment it is created. For what makes the human body human and not plant-like is precisely the fact that at each moment through the elimination of carbon the form it has taken on can be immediately destroyed and replaced by another, the carbon being united to oxygen and exhaled as carbon dioxide. As carbon would make our bodies firm and stiff like a palm tree, the breathing process wrenches it out of its stiffness unites it with oxygen and drives it outwards. Thus we gain a mobility which as human beings we must have. In plants however (and even in annuals) carbon is held fast within a fixed form.

There is an old saying that "Blood is a very special fluid". We are right in saying that the human ego pulsates in the blood and manifests itself physically in doing so; or speaking more strictly it is along the tracks provided by the carbon, in its weaving and working, forming and unforming of itself that the spiritual principle in man called the ego, moves within the blood, moistening itself with sulphur. And just as the human ego, the essential spirit of man, *carried by Hydrogen, and expressing itself through* carbon, so also does the world-ego live (through the mediation of sulphur) in that substance that is ever forming and unforming itself - carbon. The fact is that in the early stages of the Earth's development it was carbon alone which was deposited or precipitated. It was not until later that, for example, limestone came into existence, supplying man with the foundation for the creation of a more solid bony structure. In order that the organism which lives *through* the carbon might be moved about, man and the higher animals provided a supporting structure in the skeleton which is made of lime. In this way, by making mobile the carbon form within him, man raises himself from the merely immobile mineral lime formation which the earth possesses and which he incorporates in order to have solid earth-matter within his body. The bony lime structure represents the solid earth within the human body

Let me put it in this way: Underlying every living being there is a scaffolding of carbon, more or less either relatively permanent or continually fluctuating, in the tracks of which the spiritual principle, *embodied in hydrogen,* moves through the world. Let us make a schematic drawing of this so that you can see the matter quite clearly before you.

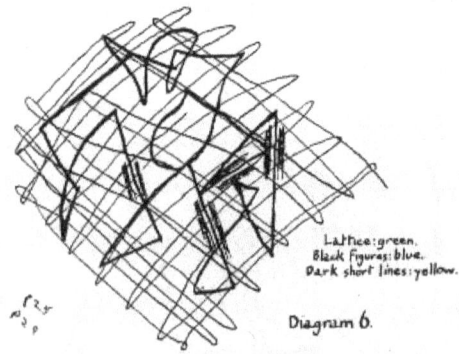

Lattice: green.
Black figures: blue.
Dark short lines: yellow.

Diagram 6.

(drawing No. 6) Here is such a scaffolding which the spirit *and hydrogen,* builds up somehow or other with the help of sulphur. Here we have either the continuously changing carbon which moves in the sulphur in highly diluted form or else we have, as in the plants a more or less solidified carbon structure which is united with other ingredients. Now as I have often pointed out, a human or any other living being must be penetrated by an etheric element which is the actual bearer of life. The carbon structure of a living being must therefore be penetrated by an etheric element, which will either remain stationary about the timbers of this scaffolding, or retain a certain mobility. But the main thing is that the etheric element is in both cases distributed along the scaffolding.

This etheric element could not abide our physical earth world, if it remained alone. It would slide through instead of gripping what it has to grip in the physical earthly world if it were without a physical bearer. For it is a peculiarity of earth conditions that the spiritual must always have physical bearers. The materialists regard the physical bearer only and overlook the spiritual. To an extent they are right, because it is indeed the physical bearer which is first met with. But they overlook the fact that it is the spiritual which makes necessary everywhere the existence of a physical bearer. The physical bearer of the *energetic activity* which works in the Etheric element (we may say that the lowest level of the spiritual works in the etheric); this physical bearer which is permeated by the etheric element and *again* "moistened" as it were with sulphur, introduces into physical existence not the form *(H),* not the structure *(C),* but a continuous mobility and vitality. This physical carrier which, with help of sulphur, brings the vital activities out of the universal ether into the body, is oxygen.

Thus the part which I have coloured green in my sketch can be regarded, from the physical point of view as oxygen, and also as the brooding vibrating etheric element which permeates it. It is in the track of oxygen that the etheric element moves, with the help of sulphur.

It is this that gives meaning to the breathing process. When we breathe we take in oxygen. When the present day materialist talks of oxygen all he means is the stuff in his test-tube when he has decomposed water through electrolysis. But in oxygen there lives the lowest order of the supersensible, the etheric element: it lives there and will not be killed, as e.g. in the air around us. In the atmosphere around us the living principle in the oxygen has been killed in order that it may not cause us to faint. For any excess of the ordinary growth forces within us, if it appears where it should not be, will cause us to faint or worse. If therefore we were surrounded by an atmosphere which contained living oxygen, we should reel about as though completely stunned by it. The oxygen around us has to be killed. And yet oxygen is from its birth the bearer of life, of the etheric element. It becomes the bearer of life as soon as it leaves the sphere in which it has the task of providing a surrounding for our human external senses. Once it has entered into us through breathing it comes alive again. The oxygen which circulates inside us is not the same as that which surrounds us externally. In us it is living oxygen, just as it also becomes living oxygen immediate it penetrates into the soil, although in this case the life in it is lower in degree than it is in our bodies. The oxygen under the earth is not the same as the oxygen above the earth. It is very difficult to come to and understanding with physicists and chemists on this subject, for according to the methods they employ the oxygen must always be separated with its connection with the soil. The oxygen they are dealing with is dead, nor can it be anything else. But every science which limits itself to the physical is liable to this error. It can only understand dead corpses. In reality oxygen is the bearer of the living ether and this living ether takes hold of the oxygen through the mediation of sulphur.

We now have pointed out two extreme polarities: On the one hand the scaffolding of carbon within which *hydrogen and* the human ego - the highest form of the spiritual given to us here on earth, displays its forces or with the case of plants the world-spiritual which is active in them. On the other hand we have the human process of breathing, represented in man by the living oxygen which carries the ether. And beneath it we have the scaffolding of carbon which in man permits of his movement. These two polarities must be brought together. The oxygen must be enabled to move along the paths marked out for it by the scaffolding: it must move along every track that may be marked out for it by the *hydrogen* carbon *complex*, by the spirit *acting with* carbon; and throughout Nature the oxygen bearing the etheric life must find the way to the carbon bearing the spiritual principle . How does it do this? What here acts as the mediator?

The mediator is nitrogen. Nitrogen, *as the carrier of the Astral,* directs the *Etheric* life into the *Physical* form which is embodied into the carbon. Wherever nitrogen occurs its function is to mediate between life *carrying oxygen* and the spiritual element *of hydrogen*, which has first been incorporated in the carbon substance. It supplies the bridge between oxygen and carbon - whether it be the animal and vegetable kingdoms, or in the soil. That spirituality which with the help of the sulphur busies itself within the nitrogen is the same as we usually refer to as astral. This spirituality, which also forms the human astral body is also active in the earths surroundings from which it works in the life of plants, animals and so on.

In man and animals, the astral body is connected with the physical body through the etheric body and a certain connection is the normal state. Sometimes, however, the connection between the astral body and the physical body (or one of the physical organs) is closer than would normally be the case; so if the etheric body does not form a proper "cushion" between them, the astral intrudes itself too strongly into the physical body. It is from this that most diseases arise.

Thus spiritually speaking we find the astral element or principle placed in between oxygen and carbon; but the astral element uses nitrogen for the purpose of revealing itself in the physical world. Wherever there is nitrogen there the astral spreads forth in activity. The etheric life-element would float about in every direction like clouds and ignore the framework provided by the carbon were it not for the powerful attraction which this framework possesses for nitrogen; wherever the lines and paths have been laid down in the carbon there nitrogen drags the oxygen along; or more strictly speaking, the astral in the nitrogen drags the etheric element along these paths. Nitrogen is the great "dragger" of the principle towards the spiritual. Nitrogen is therefore essential to the soul of man since the soul is the mediator between life. i.e. without consciousness and spirit. There is indeed something very wonderful about nitrogen. If we trace its path as it goes through the human organism we find a complete double of the human being, such a "nitrogen man" actually exists. If we could separate it from the physical we should have the most beautiful ghost imaginable for it copies in exact detail the solid shape of man. On the other hand, nitrogen flows straight back into life.

Now we have an insight into the breathing process. When he breathes man takes in oxygen, i.e. etheric life. Then comes the internal nitrogen and drags the oxygen along to wherever there is carbon i.e. to wherever there is weaving and changing form. The nitrogen brings the oxygen along with it in order that the latter may hold on the carbon and set it free. The nitrogen is thus the mediator whereby carbon becomes carbon-dioxide and as such is breathed out. Only a small part, really of our surroundings consists of nitrogen, the bearer of astral spirituality. It is of immense importance to us to have oxygen in our immediate surroundings, both by day and by night. We pay less respect to the nitrogen around us in the air which we breathe because we think we have less need of it, and yet nitrogen stands in a spiritual relation to us.

The following experiment might be made: One could enclose a man in a gas-chamber containing a given volume of air and then remove a small quantity of nitrogen , so the air would be slightly poorer in nitrogen than it normally is. If this experiment could be carefully carried out it would convince you that the necessary quantity of nitrogen is at once restored, not from outside, but from inside the man's body. Man has to give up some of his own supply of nitrogen in order to restore the quantitative condition to which the nitrogen is accustomed. As human beings it is necessary that we should maintain the right quantitative relation between our whole inner being and the nitrogen around us; the right quantity of nitrogen outside us is never allowed to become less. For the merely vegetative life of man a less quantity than the normal will do , because we do not need nitrogen for the purpose of breathing But it would not be adequate to the part it plays spiritually. For that the normal quantity of nitrogen is necessary.

This shows you how strongly nitrogen plays into the spiritual and will give you some idea of how necessary this substance is to the life of the plants. The plant growing on the ground has at first only its physical body and etheric body but no astral body; but the astral element must surround it on all sides. The plant would not flower if it were not touched from the outside by the astral element. It does not take in the astral element as do men and the animals but it needs to be touched by it from outside. The astral element is everywhere and nitrogen, the bearer of the astral, is everywhere; it hovers in the air as a dead element, the moment it enters into the soil it comes to life again. Just as oxygen comes to life when drawn into the soil, so does nitrogen. This nitrogen in the earth not only comes to life but becomes something which has a very special importance for agriculture because—paradoxical as it may seem to a mind distorted by materialism - it not only comes to life but becomes sensitive inside the earth. It literally becomes the carrier of a mysterious sensitiveness which is poured out over the whole life of the earth. Nitrogen is that which senses whether the right quantity of water is present in any given soil and experiences sympathy; when water is deficient it experiences antipathy. It experiences sympathy when for any given soil the right sort of plants are present, and so on. Thus nitrogen pours out over everything a living web of sensitive life, above all nitrogen knows all those secrets of which we know nothing in an ordinary way, of the planets Saturn Sun, Moon and so on, and their influences upon the form and life of plants, of which I told you yesterday, and in the preceding lectures. Nitrogen that is everywhere abroad, knows these secrets very well. It is not at all unconscious of what emanates from the stars and becomes active in the life of plants and of the earth. Nitrogen is the mediator which senses just as in the human nerves and senses system, it also mediates sensation. Nitrogen is in fact the bearer of sensation. Thus if we look upon nitrogen, moving about everywhere like fluctuating sensations, we shall see into the intimacies of the life in Nature. Thus we shall come to the conclusion that in the handling of nitrogen something is done which is of enormous importance for life of plants. We shall study this further in the subsequent lectures.

In the meantime there is, however, one thing more to be considered. There is a living co-operation of the *Hydrogen's* spiritual principle which has taken shape within the carbonic framework, with the astral principle working within nitrogen, which permeates that framework with life and sensations, that is, stirs up a living agility in the oxygen. But in the earthly sphere this co-operation is bought about by yet another *aspect of an* element, which links up the physical world with the expanses of the cosmos. For the earth cannot wander about the Universe as a solid entity cut off from the rest of the Universe. If the Earth did this it would be in the same position as a man who lived on a farm, but wished to remain independent of everything that grew in the fields around him. No reasonable man would do that. What today is growing in the fields around us tomorrow will be in human stomachs and later will return to the soil in some form or another. We human beings cannot isolate ourselves from our environment. We are bound up with it and belong to it as much as my little finger belongs to me. There must be a continuous interchange of substances, and this applies also to the relation between earth with all its creatures and the surrounding Cosmos. All that is living on earth in physical shape must be able to find its way back into the Cosmos where it will be in a way purified and refined.

We have in the first place the carbon framework (which I have coloured blue in the drawing), then the etheric oxygenous life-element (coloured green) and then, proceeding from the oxygen and enabled by nitrogen to

96

follow the various lines and paths within the framework, we have the astral element which forms the bridge between carbon and oxygen. I could indicate everywhere here how the nitrogen drags into the blue lines which I have indicated schematically with the green lines. But the whole of the very delicate structure which is formed in the living being must be able to disappear again. It is not the spirit which disappears, but that which the spirit *and Hydrogen* has built up in the carbon, and into which it has drawn the etheric life borne in the oxygen. It must disappear not only from the earth, but dissipate into the Cosmos. This is done by a substance which is allied as closely as possible to the physical and yet is allied as closely as possible to the spiritual: This activity is *the second face of* hydrogen. Although hydrogen is itself the most attenuated form of the physical substance, *due it carrying the basic imprint from the stars, upon which the physical body forms,* it goes still further and dissipates physical matter which, *again helped* by sulphur, floats away into that cosmic region in which matter is no longer distinguishable. One may say then: Spirit has first become physical, *carried on Hydrogen's incarnating impulse,* and lives in the body at once in its astral form and reflecting itself as Ego. There it lives physically as spirit transformed into something physical. After a time the spirit begins to feel ill at ease. It wishes to get rid of its physical form. Moistening itself once again with sulphur, *Spirit rides Hydrogens excarnating nature,* by means of which it can yield up any kind of individual structure and give itself over to the cosmic region of formless chaos, where there is no longer any determinate organisation. Hydrogen carries away all that the astral principle has taken up as form and life, and carries it out the expanses of the Cosmos, so that it can be taken up again from thence, as I have already described. Hydrogen in fact, *brings into manifestation and then* dissolves, everything.

Thus we have these 5 substances which are the immediate representatives of all that works and weaves in the realm of the living and also in the realm of the seemingly dead, which in fact is only transiently so: Sulphur. Carbon, Hydrogen, Oxygen and Nitrogen, each of these substances is inwardly related to its own particular order of *energetic* entity. They are therefore something quite different from which our modern chemistry refers to by the same names. Our chemistry speaks only of the corpses of these substances, not of the actual substances themselves. These we must learn to know as something living and sentient, and, curiously enough hydrogen, which seems the least dense of the five and has the smallest atomic weight, is the least spiritual among them.

Now consider: What are we actually doing when we meditate? (I am compelled to add this ensure that these things do not remain among the mists of spirituality) The Oriental has meditated in his own way. We in Middle and Western Europe meditate in ours. Mediation as we ought to practise it only slightly touches the breathing process; our soul is living and weaving in concentration and meditation. But all these spiritual exercises have a bodily counterpart, however subtle and intimate. In meditation, the regular rhythm of breathing, which is so closely connected with man's life, undergoes a definite if subtle change. When we meditate we always retain a little more carbon-dioxide in us than in the ordinary everyday consciousness, We do not, as in ordinary life, thrust out the whole bulk of carbon-dioxide into the atmosphere where nitrogen is everywhere around us. We hold some of it back.

Now consider: If you knock your head against some thing hard, like a table, you become conscious only of your own pain. But if you gently stroke the   surface of the table, then you will become conscious of the table. The same thing happens in meditation. It gradually develops an awareness of the nitrogen all around you. That is the real process in meditation. Everything   becomes an object of knowledge, including the life of the nitrogen around us. For nitrogen is a very learned fellow. He teaches us about the doings of Mercury, Venus, etc. because he knows or rather senses them. All these things rest upon perfectly real processes. And as I shall show in greater detail, it is at this point that the spiritual working in the soul activity, begins to have a bearing upon Agriculture. This interaction between the soul-spiritual element and that which is around us is what has particularly interested our dear friend Stegemann,

For, indeed, if a man has to do with Agriculture it is a good thing if he is able to meditate, for in this way he will make himself receptive to the manifestations of nitrogen. If one does become receptive in this way, one begins to practise Agriculture in quite a different way and spirit. One suddenly gets all kinds of new ideas; they simply come and one then has many secrets in large estates and smaller farms.

I do not wish to repeat what I said an hour ago but I can describe in another way, Take the case of a peasant who walks through his fields. The scientist regards him as unlearned and stupid. But this is not so, simply because - forgive me but I speak the truth - simply because instinctively a peasant is given to meditation. He ponders much throughout the long winter nights. He acquires a kind of spiritual knowledge, as it were, only he cannot express it. He walks through his fields and suddenly he knows something; later he tries it out. At any rate this is what I found over and over again in my youth when I lived among peasant folk. The mere intellect will not be enough, it does not lead us deep enough, For after all nature's life and weaving is so fine and delicate that the net of intellectual concepts - and this is where science has erred of recent years - has too large a mesh to catch it.

Now all these substances of which I have spoken Sulphur, Carbon, Nitrogen, Hydrogen are united in albumen. This will enable us to see more clearly into the nature of seed formation. Whenever carbon, hydrogen and nitrogen are present in leaf, blossom, calyx or root they are always united to other substances in some form or other. They are dependent upon these other substances. There are only two ways in which they can become independent. One is when the hydrogen carries all individual substances out into the expanses of the Cosmos and dissolves them into the general chaos; and the other is when the hydrogen, *inspires the Cosmic Forces* to drive the basic element of the protein (for albumen) into the seed formation and there makes them independent of each other so that they become receptive of the influences of the Cosmos. In tiny seed there is chaos, and in the wide periphery of the Cosmos there is another chaos, and whenever the chaos at the periphery works upon the chaos within the seed, new life comes into being.

Now look how these so-called substances, which are really bearers of spirit, work in the realm of Nature. Again we may say that the oxygen and nitrogen inside man's body behave themselves in an ordinary way, for within man's body they manifest their normal qualities. Ordinary science ignores it because the process is hidden. But the ultimate products of carbon and hydrogen cannot behave in so normal a fashion as do oxygen and nitrogen. Let us take carbon first. When the carbon, active in the plant realm enters the realms of animals and man it must become mobile - at least transiently. And in order to build up the fixed shape of the organism it must attach itself to an underlying framework. This is provided on the one hand by our deeply laid skeleton consisting of limestone, and on the other hand by the silicious element which we always carry in our bodies; so that both in man and in the animals carbon to a certain extent masks *hydrogen's* formative force. It climbs up, as it were along the lines of formative forces of limestone and silicon. Limestone endows it with the earthly formative power, silicon with the cosmic. In man and the animals carbon does not as it were claim sole authority for itself, but adheres to what is formed by lime and silicon.

But lime and silicon are also the basis of the growth of plants. We must therefore learn to know the activities of carbon in the breathing, digestive and circulatory processes of man in relation to his bony and silicious structure - as though we could, as it were, creep into the body and see how the formative force of carbon in the circulation radiates into the limestone and silicon. And we must unfold this same kind of vision when we look upon a piece of ground covered with flowers having limestone and silicon beneath them. Into man we cannot creep; but here at any rate we can see what is going on. Here we can develop the necessary knowledge. We can see how the oxygen element is caught up by the nitrogen element and carried down into the carbon element, but only in so far as the latter adheres to the lime and silicon structure. We can even say that carbon is only the mediator. Or we can say that what lives in the environment is kindled to life in oxygen and must be carried into the earth by means of nitrogen, where it can follow the form provided by the limestone and silicon. Those who have any sensitiveness for these things can observe this process at work most wonderfully in all the papilionaceous plants (Leguminosae) that is in all the plants which in Agriculture may be called collectors of nitrogen, and whose special function it is to attract nitrogen and hand it on to what lies below them. For down in the earth under those leguminosae there is something that thirsts for nitrogen as the lungs of man thirst for oxygen - and that is lime. It is a necessity for the lime under the earth that it should breathe in nitrogen just as the human lungs need oxygen. And in the papilionaceous plants a process takes place similar to that which is carried out by the epithelium issue in our lungs lining the bronchial tubes. There is a kind of in-breathing which leads nitrogen down. And these are the only plants that do this. All other plants are closer to exhalation. Thus the whole organism of the plant-world is divided into two when we look at the nitrogen-breathing. All papilionacae are, as it were, the air passages. Other plants represent the other organs in which breathing goes in a more secret way and whose real task is to fulfil some function. We must learn to look upon each species of plant as placed within a great whole, the organism of the plant-world, just as each human organ is placed within the whole human organism. We must come to regard the different plants as part of a great whole, then we shall see the immense importance or these Papilionacae. True, science knows something of this already but it is necessary that we should gain knowledge of them from these spiritual foundations, otherwise there is a danger as tradition fades more and more during the decades, that we shall stray into false paths in applying scientific knowledge. We can see how these papilionacae actually function. They have all the characteristics of keeping their fruit process which in other plants tends to be higher up in the region of their leaves. They all want to bear fruit before they have flowered. The reason is that these plants develop the process allied to nitrogen far nearer to the earth (they stimulate bacteria to fix nitrogen along their roots) than do the other plants, which unfold this process at a greater distance from the surface of the earth. These plants have also the tendency to colour the leaves, not with the ordinary green, but with a rather darker shade. The actual fruit, moreover undergoes a kind of atrophy, the seed remains capable of germination for a short time only and then becomes barren. Indeed, these plants are so organised as to bring to special perfection what the plant-world receives from Winter and not from Summer. They have, therefore a tendency to wait for Winter. They want to wait with what they are developing for the Winter. Their growth is delayed when they have a sufficient supply of what they need, namely, nitrogen from the air which they can convey below in their own manner. In this way one can get insight into the becoming and living *that* goes in and *is* above the soil.

If in addition you take into account the fact that lime has a wonderful relationship with the world of human desires you will see how alive and organic the whole thing becomes. In its elemental form as calcium lime is never at rest; it seeks and experiences itself; it tries to become quick-lime, i.e. to unite with oxygen. But even then it is not content; it longs to absorb the whole range of metallic acids, even including bitumen which is not really a mineral. Hidden in the earth lime develops the longing to attract everything to itself. It develops in the soil what is almost a desire nature. It is possible if one has the right feeling in these matters, to sense the difference between it and other substances. Lime fairly sucks one dry. One feels that it has a thoroughly greedy nature and that wherever it is it seeks to draw to itself also the plant-element. For indeed everything that limestone wants lives in plants, and it must continually turn away from the lime. What does this? It is done by the supremely aristocratic element which asks for nothing but relies upon itself. For there is such an aristocratic substance. It is silicon. People are   mistaken in thinking that silicon is only present where it shows its firm rock-like outline. Silicon is distributed everywhere in homeopathic doses. It is at rest and makes no claim on anything else. Lime lays claim to everything, silicon to nothing. Silicon thus resembles our sense-organs which do not perceive themselves but which perceive the external world. Silicon is the general external sense organ of the earth, lime the representing general which desires; *the* clay *humus complex* mediates between the two. Clay —*aluminium silicate*— is slightly closer to silicon and yet *with humus* it acts as a mediator with lime. Now one should understand this in order to acquire a knowledge supported by feeling. One should feel about lime that it is a fellow full of desires, who wants to grab things for himself; and about silicon that it is a very superior aristocrat who becomes what the lime has grabbed, carries it up into the atmosphere, and develops the plant-forms. There dwells the silicon, either entrenched in his moated castle, as in the horse-tail (equisetum), or distributed everywhere in fine homeopathic doses, where he endeavours to take away what the lime has attached. Once again we realise that we are in the presence of an extremely subtle process of Nature.

Carbon, *by enacting the hydrogens archetypal impulse,* is the really *physical* formative element in all plants: it builds up the framework. But in the course of the earth's development its task has been rendered more difficult. Carbon could give form to all plants as long as there were water below it. Then everything would have grown. But since a certain period, lime has been formed underneath and lime disturbs the work, and because the opposition of the limestone had to be overcome, carbon allies itself to *its periodic table brother* silicon, and both together, in combination with clay, - *aluminium silicate, and aluminium's brother Boron* — they once again start on their formative work. How, in the midst of all this, does the life of a plant go on? Below is the limestone trying to seize it with its tentacles, above is the silicon which wants to make it as long and thin as the tenuous water-plants. But in the midst of them is carbon which *builds* the actual plant-forms and brings order into everything. And just as our astral body brings about a balance between our Ego and etheric body, so nitrogen works in between, as the astral element.

This is what we must learn to understand - how nitrogen, *stabilised within humus,* manages things between lime, clay and silicon. And also between what the lime is always longing for below, and what silicon seeks always to radiate upwards. In this way the practical question arises: What is the correct way of introducing nitrogen into the plant-world? This is the question that will lead us over to deal with the different methods of manuring the ground.

The Physical Formative Forces

GA in Italics

Foreword

*The question of the **Physical Formative Forces (PFF)** is significant as it takes up a good deal of the discussion, in the Agriculture Course. It is the Earthly and Cosmic processes that appear throughout, in relationship to Calcium and Silica. The Earthly and Cosmic activities are often seen as a twofold process of expansion and contraction. How many people take the next step, to see how each of the Cosmic and Earthly processes has two sides to their story, thus making it a fourfold tale?*

This fourfold story is told most clearly in the 8th lecture of the Agriculture Course , however it starts in lecture 1 and becomes 4 fold in the 2nd lecture. Another important step occurs in the 6th lecture. If the PFF story is seen in 2nd lecture, then the subsequent lectures take on a very different interpretation, to what we commonly find. The biggest highlight is we are given 4 more preparations with which to influence nature.

As we see in the 2nd lecture RS places this overall story within the context of the 3 fold physical body, organised into a nerve sense system, a rhythmic system and a metabolic system, and then within each of these we have cosmic and earthly processes occurring.

While this story is in the Agriculture course I have found other lectures where RS enlarged on these images. He appears to have spoken of these activities many times. My goal is to present Dr Steiner in his own words, with my editing in italics helping the story to make sense, by using a coherent language throughout. My edits are most often naming the force being described in the Cosmic and Earthly language.

*Most of the diagrams I provide, are orientated towards the northern hemisphere Zenith and are thus best viewed facing **South**. There are also some diagrams here orientated to the Earth's **North** magnetic pole. Nature is very magnetic.*

The Three Worlds

3 fold Physical Body & 4 Physical Activities

We find above all that when through Imagination and Inspiration we enter the spiritual world in full consciousness, it immediately appears to us to be threefold. Hence we can speak of the world, and of our theme, the evolution of the world and of man, only when we have come to the point we have now reached. Only now can I speak of how a man, confronted by the external world, by all that manifests itself to the senses, is really facing the spiritual world in its threefold nature — facing actually three worlds. Once the veil has been lifted which creates the chaos, we no longer have one world only before us, but three worlds, and each of the three has its definite connection with the human being.

When we succeed in penetrating this veil of chaos — later I shall be showing how we can also describe this as crossing the threshold of the spiritual world — we perceive the three worlds. The first of the three is really the world we have just left, somewhat transformed but still there for spiritual existence. When the veil of chaos has been thrust aside, this world appears as though it were a memory. We have passed over into the spiritual world; and just as here

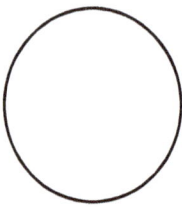

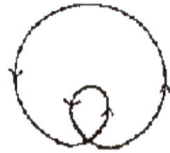

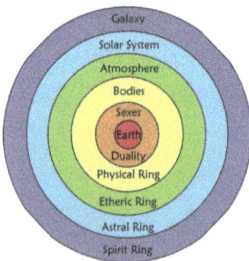
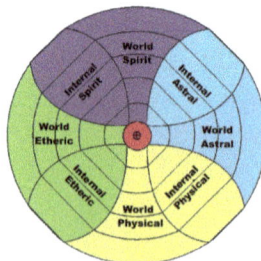
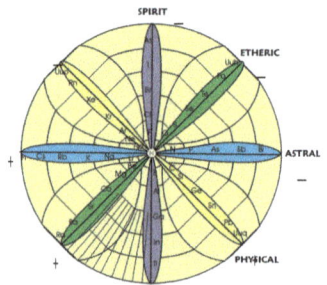

Stage 1	Stage 2	Stage 3
Archetype	Forming	Manifestation
Thesis	Antithesis	Synthesis
Passive	Pulsating	Enfolded
The Constant Field	Surrounding the form	Form
Spirit	Soul	Body
Saturn	Jupiter	

Sout

we remember certain things, so in the spiritual world we remember what constitutes the physical world of the senses. Here, then, is the first of the three worlds.

The second world we encounter is the one I have called in my book, Theosophy, the soul-world.

And the third world, the highest of the three, is the true spiritual world, the world of the spirit.

To begin with, I shall give you only a schematic account of all this, but from the way these three worlds are related to man you will gather many things about them. To these three worlds as they appear in three ascending stages — the lowest, the middle one, and the highest — I will then relate man's three members — the head; then the breast-organisation embracing all that is rhythmical, the breathing system and blood circulation; thirdly, the metabolic-limb system, which includes nutrition, digestion and the distribution throughout the body of the products of digestion, all of which engender movement. All this has to do with the metabolic-limb system. If this scheme were drawn, there would have to be a closed circle for the breast; for the head a circle left open, and open also for the limb system. When perceived physically, man's head appears to be closed above and would have to be drawn so, but perceived spiritually, it is open. The part of a man which does not belong at all to the realm of the spirit is the bony system, which is entirely of a physical nature; and when spiritually you study the human head, its thick skull is not seen. Only the skin is visible where the hair grows.

When this is looked at spiritually, however, something else appears. Ordinary hair is not there at all, but purely *astral* hair; in other words, *astral* rays which penetrate into the human organism and are held back, to some extent, only by the physical hair. But it is just where there is bone in the organism that the spirit can enter most easily, and this it does in the form of rays. So, on first looking at a man with your physical eyes, you see his physical form with the head above, and on his head — if he is not already bald — there is hair. But then, where the dome of the skull comes, spiritually you see nothing of the physical man; you see *Spirit filled Cosmic*

Force rays, sun-like rays, pouring into him from the spiritual worlds.

Thus the reason for the circle not being closed for the head is that the surrounding bony vault of the skull enables the spirit and *Cosmic Forces* to have continual access there. Nothing in a man is without purpose. By deliberate intent of the ruling powers — one might say — he has been given a head thus closed above, for here the spirit has the easiest access to his inner being because of the very thickness of the bone.

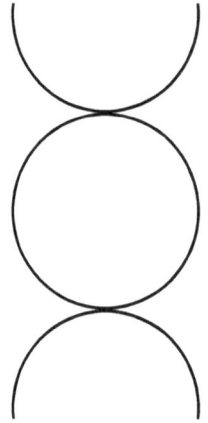

When we are in a position to observe man spiritually, we are astonished to discover how empty his head is of anything drawn from his own inner being. As regards the spiritual, he has almost nothing in him to fill the hollow globe sitting on his shoulders. Everything spiritual has to enter it from outside, *as Cosmic Forces.*

It is not thus with the other members of the human organism; as we shall soon hear, these are by their very nature *energetic.* We can distinguish in man three members — head, or nerves and senses system, rhythmic system, metabolic-limb system, and they have a quite definite relation to the three worlds: the physical world, the soul-world, and the spiritual world. I will now go further into this.

On the one hand we have to do with a threefold nature of the *Astral bodies* soul being: in forming mental images, in feeling, and in will impulses. This threefold nature of the soul being, however, corresponds very precisely with a threefolding of the physical-bodily being: a kind of head system or nerve-sense system, a rhythmic system, and a metabolic-limb system. I must stress particularly that this constitution of the human organism must not be understood merely intellectually but through inner perception. A person would be unable to comprehend how matters actually stood if he remained with an external picture, if he understood the head system as something that simply ends at the neck, the circulatory or rhythmic system as being encompassed by the trunk, while the digestive system encompasses the limb system, the sexual system.

What is important here is that while the nerve-sense system is located primarily in the head, it nevertheless extends over the entire remaining organism as such. We may thus say that when we speak here with an anthroposophical purpose about the nerve-sense system, it is the system of functions in the human organism (for we are concerned here not with spatial limitations but functional limitations) that is located essentially in the head; nevertheless the head activity extends over the entire human being so that in a certain sense the whole human being is head. The same is true for the other systems. *We must gain an* understanding for how the threefold constitution of the human being is functional, and not defined by spatial limitations.

When an individual really understands this constitution of the human being — about which many lectures could be given to describe it in full detail — he reaches the point of being able to perceive clearly the distinctions between the head system, and therefore the nerve-sense system, on the one hand, the metabolic-limb system on the other hand, and the mediating system, the rhythmic system, whose essential role is to bring about the balance between the two other systems.

If we thus wish to encompass the entire nature of the human being, we must consider the following. The actual conceptual and perceptual activity of the human being has as its basis — one cannot even say as its tool, but as its physical basis — everything that takes place physically in the nerve-

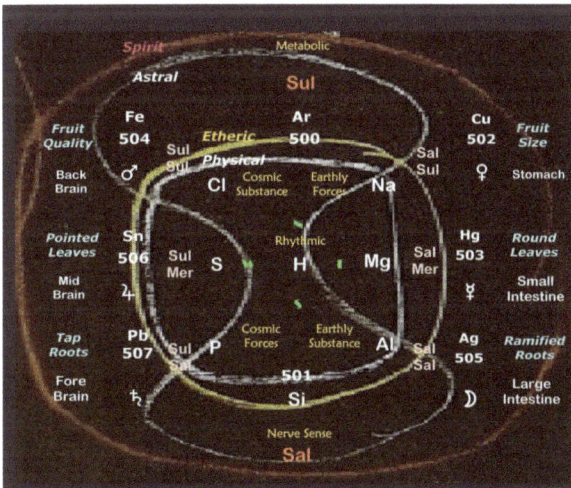

Earth and Plants   **Humans**

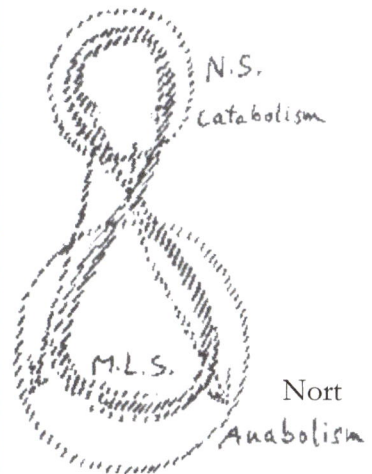

sense system. It is not the case, as is suggested by modern psychology and physiology, that those processes connected primarily with the feeling and willing systems also take place in the nerve-sense system. Such an opinion does not hold up before a more precise study of the issue. You will find such a precise study, at least suggested in its outlines, in my book, Riddles of the Soul.

Much detailed work must still be done in this regard, however. Then what spiritual science has to say with certainty from its side will be elaborated from the other side, from the physical-empirical side. It will become clear that man's feeling is not connected in a primary way with the nerve-sense system but with the rhythmic system, that just as the nerve-sense system corresponds to mentally active perception, so the rhythmic system corresponds to feeling. Only through the interaction of the rhythmic system with the nerve-sense system, by the roundabout route of the rhythm in the cerebral fluid, pulsating against the nerve-sense system, is the nerve-sense system engaged as the carrier of the conceptual life. Then, if we raise our feelings to mental images, the dull, dreamlike life of feelings is perceived and pictured by us in an inner way. Just as the life of feeling is directly connected with the rhythmic system and is indirectly mediated by it, so the life of will is connected directly with the metabolic system. This connection in turn acts in a secondary way, since metabolism takes place also in the brain, of course, so that the metabolic system in its functions presses against the nerve-sense system. In this way we are able to bring forth the mental images of our will impulses, which otherwise would unfold in a dull sleep-life within our organism.

Thus you can see that in the human organism we have three different systems that carry the *Astral bodies'* soul life in different ways. These systems do not simply differ from one another; they actually oppose each other (as I said, I can only sketch these matters today) so that on one side we have the nerve-sense system and on the other side all that constitutes the functions of the metabolic system, the metabolic-limb system (see drawing). Regarding the connection of the metabolism with the limbs, you can arrive at appropriate images if you simply consider the influence of the moving limbs on the metabolism. This influence is much greater than is ordinarily assumed in outer consciousness.

These two systems, however, the nerve-sense system and the metabolic-limb system, are in opposition, are polar opposites in a certain way. This polar opposition must be studied carefully in order to arrive at a sound pathology and therapy, particularly a pathology that could lead organically over into therapy; it must be studied carefully in all its countless individual details. If one enters into the detailed effects, it becomes evident that what I suggested yesterday is truly the case.

Within everything connected with the head system or nerve-sense system, we have breakdown processes, so that while our conceptual activity takes place in the waking state, when we perceive and form mental images, this activity is not bound up with growth and upbuilding processes but with breakdown processes, processes of elimination. This can be grasped if one

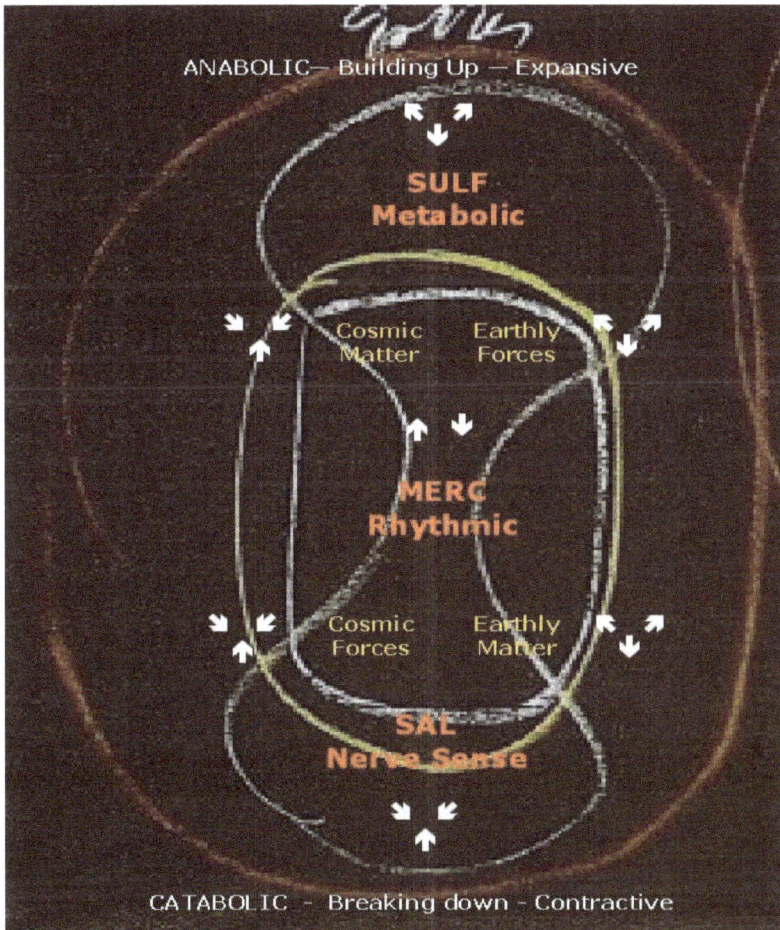

ANABOLIC— Building Up — Expansive

SULF
Metabolic

Cosmic Matter   Earthly Forces

MERC
Rhythmic

Cosmic Forces   Earthly Matter

SAL
Nerve Sense

CATABOLIC - Breaking down - Contractive

looks in a sound way at what empirical-physiological science has already presented concerning this. There is already empirical evidence or to express it better, empirical corroboration — for what spiritual science provides through its perception. You need only pursue what certain inspired physiologists are able to present about the physical processes in the nervous system, which unfold as parallel phenomena to perceiving and forming mental images. You will see then that this assertion is certainly well supported, the assertion that when we think, when we think and perceive wakefully, we have to do with processes of elimination and breakdown, not with upbuilding processes. By contrast, where the will processes are mediated for the human being in the metabolic-limb system we are concerned with upbuilding processes.

All individual functions in the human being definitely interact with one another, however. If we look at the matter correctly, we must say that the upbuilding processes from below work up into the breakdown processes, and that the breakdown processes from above work down into the upbuilding processes. Then if you pursue this logically you have the rhythmic processes as a balancing system, as functions introducing the balance between the upbuilding processes and the breakdown processes, rhythmic processes that press breakdown into build-up and build-up into breakdown.

If we do not study the matter purely outwardly, we see that in the so-called blood circulation of the heart, in the aeration of the human body, we have everywhere special processes, as it were, that are somehow interrupted. I cannot go further now into this interruption, which has its purpose, but everywhere we have a specialization of this rhythmic curve that I have sketched here. The course of breathing is a special aspect of this curve, the process that you draw if you follow the course of the blood from the heart upward toward the head or respectively toward the lungs and down to the rest of the body. Thus you have a specialization of these processes. In short, if you enliven what is suggested here, you penetrate into the functional tissue of the human organism, not in the dead way that is customary but in a living way. To do so you must enliven your own mental images. A mobile image of the human organism can thus be pictured. The human organism cannot be encompassed with static, abstract mental images, as modern physiology and pathology would like to encompass it today; it must instead be grasped with mental images in movement, with mental images that can really penetrate into the working of something that has inner movement, that is in no way merely a mechanical interaction of organs situated at rest in relation to one another.

We thus can see that within the human organism there is basically a continuous interaction between the breakdown processes, the deadening processes, and the upbuilding processes, the growth or proliferative processes. The human organization cannot be grasped without this activity.

What is actually present there, however? Let's look at the matter more precisely. If the breakdown process of the nerve-sense organization works into the metabolic-limb system through rhythm, something is present there that works against the metabolic-limb system, something that is a poison for this metabolic-limb system. The reverse is also the case, that what is present in the upbuilding system, working into the head system in rhythm, is a poison for the head system. And since, as I have indicated, the systems are spread out over the entire organism, a poisoning and unpoisoning are continuously taking place everywhere in the human organism, and this is brought into balance by the rhythmic processes.

We are therefore unable to regard such a natural process as taking its course one-sidedly, in the way that one normally pictures things, so that healthy processes are simply designated as normal. Rather we look into two processes working against one another, where one is a process that is thoroughly illness-engendering for the other. We simply cannot live in the physical organism at all without continuously exposing our metabolic-limb system to the causes of illness from the head system and exposing the head system to the causes of illness from the metabolic system. A scale that is not balanced properly is thrown out of balance by entirely natural laws so that the beam does not rest on the horizontal; similarly life, because it is in constant movement within itself, does not simply exist in a state of balanced rest but rather exists in a state of balance that can deviate in both directions toward irregularities.

Healing, then, means simply that if the head system, for example, is working in a way too strongly poisonous on the metabolic system, its poisoning effect is relieved, its poisonous effect is taken away. If, on the other hand, the metabolic-limb system is working in a way too strongly poisonous on the head system, which means working over abundantly: then its poisonous effect must also be removed.

It is possible to arrive at a comprehensive view of this realm, however, only if one now extends what can be observed in the human being to the observation of all nature, if one is able to grasp all nature in a spiritual scientific sense. If you look at the plant-forming process, for example, you can see clearly and macroscopically the upward striving of plant-forming processes, a striving away from the center of the earth. You may make a stimulating study of this metamorphosing formative striving of the plants, at least in a rudimentary way, on the basis of the guidelines offered in Goethe's Metamorphosis of the Plants. In Goethe's Metamorphosis of the Plants there is a sketchy rendering of the first composition, the first elements that are to be studied about the nature of the plant in this direction, but the direction of such a study must be developed further. The initial guidelines must be pursued, for then we may obtain a living view of everything involved in plant growth: when rooting in the soil the plant's *Cosmic Forces* upward-striving develops in a negative direction in the root; the plant begins to grow, then grows upward with the *Cosmic Forces*, overcoming the force of attraction of the *Earthly Substance* prevailing in the root; then it wrestles through the *Earthly Forces* in order to come ultimately to *the Cosmic Substance's* blossom, fruit, and seed formation. A great deal takes place upon this path.

On this path, for example, an opposing force once again intervenes. The opposing force that intervenes can be well observed if you study, simply to take an example, the common birch, betula alba. Pursue very precisely the process that takes place from the root formation through the trunk formation, particularly the bark formation. Consider how, on the basis of everything that works together in the trunk and bark formation, there develops what later comes into manifestation in the leaf formation. This can be studied particularly well in a spiritual scientific way if the still-brownish young birch leaves are studied in the spring.

If this is studied vividly, one also receives a view of forces self-metamorphosing, forces that are active there within the plant. One receives a view of how, on the one hand, there is a formative *Cosmic* force active in the process of plant formation that works from below upward. On the other hand it is also possible to behold the *Earthly Substance* that retards, which in the root still, works strongly as the force of gravity but which, as the plant wrestles itself free from the earthly substance, out into the air is able to work together in another way with the *Cosmic Force* upward-striving force. We then reach an interesting stage, a stage very helpful in understanding how in plant formation during this upward-striving process certain salts, potassium salts, are deposited in the birch bark; this is simply the result of the interaction of the *Cosmic Substance* working downward with the *Cosmic* forces working upward, tending toward protein-formation, you could say, toward what I would like to designate as the albuminizing force formation.

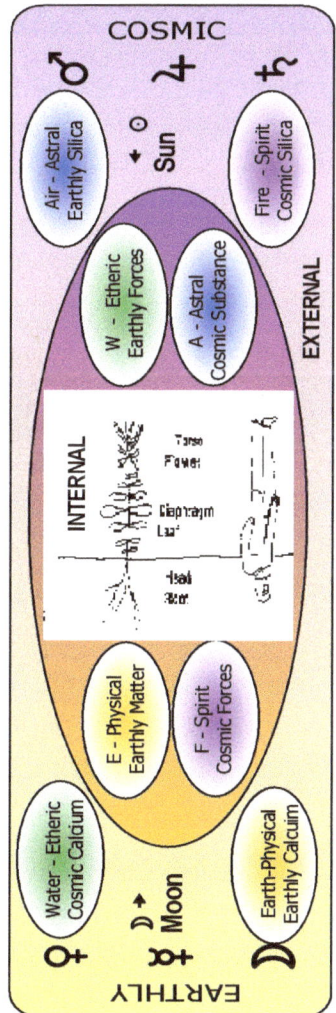

In this way it is possible to penetrate into the plant-forming process. I can only indicate this here. By looking at how the potassium salts are deposited in the birch bark, how something wrestles itself free from this *Cosmic Substance*, drawing downward (a process somewhat comparable to what happens when a salt precipitates out of a solution), coming to the process that takes place when the solution rids itself of the salt, we come to see, to grasp in a living way, the process of protein formation, the process I would designate as the albuminizing process. We thus have a path to study what outwardly surrounds the human being, to study it vividly.

Then when we look back at the human being, we can see how, fundamentally speaking, the human being has the same form of forces in him — if we consider the *Cosmic Force* breakdown process working from above downward — that work from below upward in the plant. We can see that in what is active in the *Cosmic* forces working downward from the head system toward the metabolic-limb system there is something like an inverted plant element active within us. We can see that in fact those *Cosmic* forces that we see sent upward in plant growth work in a downward direction in the human being. If the human being inappropriately holds back this *Cosmic Force* process of plant formation active within him, so that he doesn't permeate the bodily life in the right way with *the Cosmic Forces* **active in the head — the astral, the ego-being —** and if this then penetrates the bodily nature, this penetration expressing itself within the body, then *the Spirit* is held up there, something that should proceed into the human organism. We thus have to do with a pathological phenomenon like that which confronts us, for example, in cases of rheumatism or gouty conditions. If we study what is brought about in the human organism when this breakdown process is dammed up in a certain way, we discover its effects in the process of rheumatism, in the process of gout formation, and so on.

Let us now shift our gaze again from within the organism to a process of plant formation like the one we have in the betula alba. From this we can arrive at the following. We look on the one hand into what takes place in salt formation and on the other hand into protein formation. We find, if we understand this process of protein formation in the right way, that the opposite process is within it and is held up there. We find held up in the organism that process which should take place in a way similar to the correct process of albuminizing in the leaves of the birch. We are thus able to come to the relationship between those processes that take place in the birch leaves, for example, and the processes within the organism if we process what is in the birch leaves into remedies. We can then give these remedies to the human being, by means of which we can bring about a healing, because the remedy correctly opposes this damming-up process that occurs in rheumatism and gout. *The Spirit is drawn into the metabolism, providing direction to the astral activity there.* In this way we look both at what is taking place outside in nature and at what takes place within the organism, and then we arrive at an idea of how we should guide the healing forces.

On the other hand we can see instances when the *Cosmic Force* breakdown processes proceed in such a way that the organism cannot restrain them, so that they pour themselves downward, and the rhythmic system does not press them back in the right way; they thus reach the periphery of the body pressing outward, as it were, toward the skin. Then we get inflammatory conditions on the outer portion of the human being, we get skin eruptions and the like. If we now look hack again to our plant, to the *betula alba*, we find the opposing process in the disposition of the potassium salts in the birch bark: we thus become able to see how we can fight against the process of skin eruption, which is an excessive function of exudation within the human being, by preparing a remedy from the birch bark.

We are therefore able to study how plant processes, how mineral processes, are active, and we grasp the connections between what is outside in nature and what is active within the human being. In other words, medical empiricism, therapeutic empiricism, ascends to what Goethe calls in his sense — not now in the intellectual sense but in his sense the rational stage of science. We arrive at a science as therapy, which is able really to penetrate into the connections.

These things are not so simple, for one must study things in detail, at least in accordance with certain types, at first in accordance with secret types of the human personality and in accordance with secrets of natural existence. It should not be assumed that if the process has been studied in an example such as the betula alba, an overview has already been reached of what needs to be considered. In each different plant-forming process — for example in the horse chestnut or whatever — these formative processes will manifest themselves in an essentially different way. What has been indicated here should not in any way lead to a generalized twaddle but to a very serious and extensive study. (7)

First of all, it will be well to distinguish, in each of the three worlds, **substance** from **activity.** In reality, substance and *force* are one, but they work in different ways in the world. You gain a clear idea of this from the substance of your own being. You have substance in your arm, and when this substance is out of order you will feel pain of some kind; it is obvious that something within the substance of the arm has gone wrong. If the *force* of the arm is not properly controlled, you may perhaps hit your neighbour and he feels pain. This shows that the *force* is out of gear. Nevertheless, though manifesting outwardly in different ways, the substance and *force* in your arm are one.

If now we turn to the human head, we find its *Earthly* substance derived entirely from the physical world. During the formation of the human embryo the *Earthly* substance of the head comes from the parents; and the subsequent development of the head, and of the whole head and nerve-senses system, depends for its substance entirely on the earthly-material world. On the other hand, all the activity that has to do with the plastic forming of a man's head, the activity by means of which its *Earthly* substance is given *shape*, comes entirely from the spiritual world as *Cosmic Forces*. So that in respect of activity, the head is entirely a *Cosmic* formation. Therefore the head has to be left open — in a spiritual sense — so that *Cosmic Forces* can play into it.

At any time of life you can thus say: The substance of my head

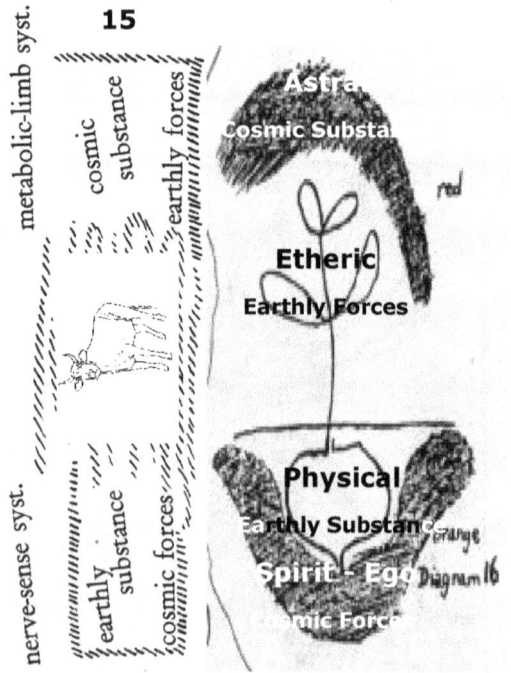

15

Diagram 16

page labels: metabolic-limb syst. · cosmic substance · earthly forces · Astral · Cosmic Substance · red · Etheric · Earthly Forces · Physical · Earthly Substance · orange · Spirit - Ego · Cosmic Force · nerve-sense syst. · earthly substance · cosmic forces comes entirely from the Earth, but it is put together and plastically formed in such a way that it cannot be the work of earthly forces. The forms of this human head are shaped entirely from the spiritual world; *by Cosmic Forces*, they might be called a heavenly creation. Anyone who contemplates spiritually the human head, in relation to the world, has to go far and deep.

Now in the same way he turns his gaze to a plant. He says to himself: The plant has a definite form *from the Stars*. Its substance is drawn from the earth, but its form comes *via* the etheric world — hence still from the spatial world.

Then he looks at an animal. The animal — he will say to himself — derives the *Earthly* substance of its head entirely from the world of space, but something *Cosmic* certainly flows into its *Force* activity.

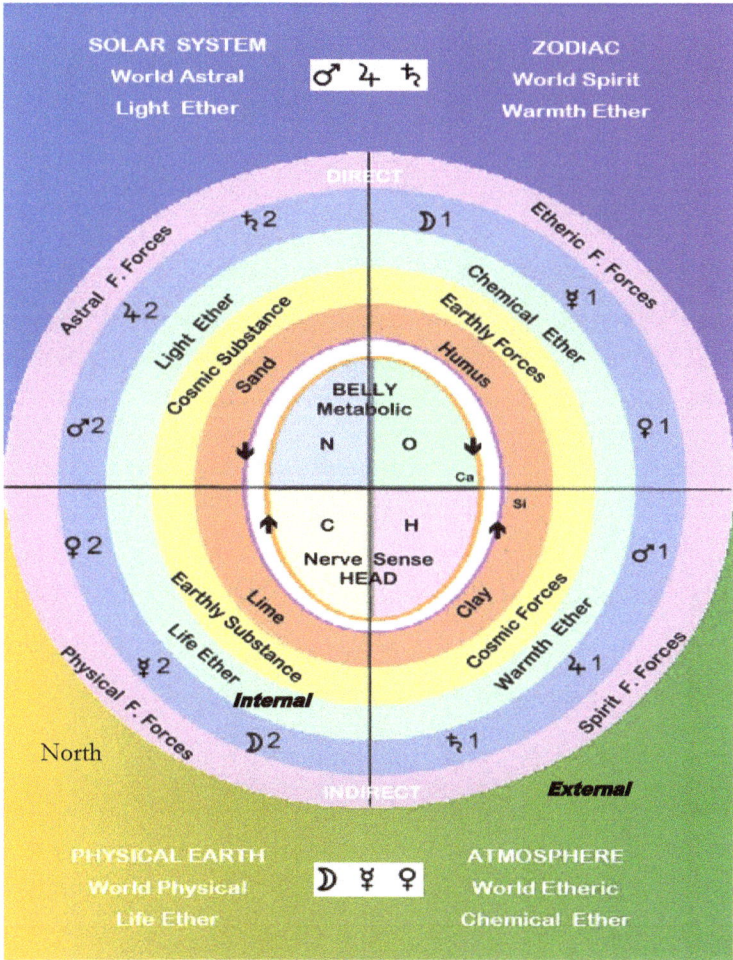

When we come to the human head, however, we find for the first time that something of the highest spirituality, something that can be called heavenly, is playing in. We see that the human head could never arise from earthly forces, though its substance is taken from earthly materials. So in the human head, which is itself a kind of miniature Cosmos, the spiritual world builds up a *Cosmic force* form out of earthly substance.

It is precisely the reverse with the metabolic-limb system, which embraces the organs for external movement — legs, arms — and the extension of these within the body — the digestive system. *There we see Earthly Forces directing Cosmic Substance.*

For the present I am leaving out the middle system — the rhythmical system which embraces breathing and the circulation of the blood. I will deal now with the system which brings together the processes of digestion and nourishment, and the inner combustion which enables a man to move.

Now the substance of this metabolic-limb system is not derived from the Earth. Improbable as it may sound, you bear within your metabolic-limb man something which is not of earthly origin but consists wholly of *Cosmic* substance from the third world, the world of the spirit. You may say: But I can see my legs; they are physically perceptible, which they would not be if they consisted of *Cosmic* substance. This objection is quite justified, but there is something more to be considered.

Your real legs are indeed *cosmic* throughout; your real arms too; but the material for them is provided by your head. The head is the organ which fills spirit arms, spirit hands, spirit legs, spirit feet, with substance; and this *Cosmic* substance penetrates into the spirituality of the limbs and of the digestive organs. So that something which in reality belongs entirely to the spiritual world is permeated, flooded, with physical matter by the head. That is why it is so difficult to grasp with the ideas of physical science that a man consists of head-breast-limbs-digestive organs. People think of the head as being there at the top, and they assume that when a man is decapitated he has no head left. It is not so, however; a man is substantially head all over. Even right to the end of his big toe he is head, for his head sends down its substance there. It is only the *Earthly* substance of the head that is earthly in origin, and the head gives its earthly-material character to the other substances; while the *Cosmic* substance of the metabolic-limb organs comes from the spiritual world.

If through vigorous auto-suggestion of a negative kind we can suggest away the head of a man, so that in appearance he is headless, and if we can do this not only in thought but so that we really see the man as headless, then the rest of his organism also disappears; with the head goes the whole of the man as a being perceptible to the senses. And if the head is then to be there for us at all, the rest of the man has to be perceived spiritually. For in reality we go about under the imprint of higher worlds, with spirit legs, spirit arms, and it is only the head that fills them with physical matter.

On the other hand the *Earthly* forces, the activity, for all that makes up the metabolic-limb man are drawn from the physical world. If you make a step forward or lift an arm, the mechanism involved, and even the chemical processes that take place in moving an arm or leg, or the chemical processes in the digestive organs — all this activity is the earthly *forces*. So that in your limbs you bear invisible *Cosmic* substance, but *Earthly* forces drawn from earthly life. Hence we are built up as regards our head and its substance out of the Earth, but this same head is permeated with heavenly forces. In our limbs we are built up entirely from heavenly substance; but the forces playing into this heavenly substance during our life on Earth are earthly forces — gravitation and other physical and chemical forces all belonging to the Earth.

You see, therefore, that head and limbs are opposites. The head consists of earthly *substance* and is given plastic form by *cosmic forces*. The limbs and the digestive system are formed wholly of *cosmic* substance, and would not be visible were they not saturated with earthly substance by the head. But when anyone walks, or grasps something, or digests food, the *cosmic* substance makes use of earthly forces in order that life on Earth, from birth to death, may be carried on.

In this complicated way does a man stand in relation to the three worlds.

The spiritual world participates with its activity in the head; with its *Cosmic* substance it participates in a man's third organisation, his metabolic-limb system. The lowest world, the world most dominated by the senses, participates through its *Cosmic Substance* activity in the metabolism and the movement of the limbs, and through its *Earthly* substance in the head; whereas the substance in a man's third system is wholly *Cosmic*.

In the middle system, which embraces the breathing and the circulation of the blood, spiritual activity and material substance work into each other. The spiritual activity, flowing through the movement of our breathing and the beating of our heart, is always accompanied to some extent by substantiality. And, in the same way, the substantiality of earthly existence, inasmuch as oxygen streams into the breathing, is to some extent accompanied by earthly activity. So you see that in the middle man, in man's second system, everything flows together — *Cosmic* substance and Cosmic Forces flow in here; earthly *forces* and *earthly* substance flow in there. By this means we are made receptive both to the activity of the middle world and to its substantiality.

So in this middle man there is a great deal of intermingling and for this reason we need our wonderfully perfect rhythmical system — the rhythm of the heart, the rhythm of the lungs in breathing. All the intermingling of *force* and substance is balanced, harmonised, melodised, through these rhythms, and this can happen because man is organised for it.

In the head system and the limb system, *force* and substantiality come from quite different sources, but in the middle system they come from all three worlds and in a variety of ways — at one place *force* accompanied by substance, in another place substance accompanied by *force*; here pure activity, there pure substance — all these variations flow through the middle man. If as a doctor you take a man's pulse, you can really feel there the balancing of the heavenly nature of the soul against earthly activity and substantiality. Again, if you observe the breathing, you can feel a man's inner striving for balance between the various agencies which relate him to the middle world.

All this is very complicated, you will say. It is true that a lecture-course is generally easy to understand up to a certain stage, but when it comes to the point where man's relation to the world has to be grasped, people often say: "This is becoming very difficult — we can't keep up with it." (8)

We have tried, again from a particular aspect, to place the human being into the universe. Today we wish to put the subject forward in a way which will, as it were, weld everything into a whole. During our physical life we live upon the earth; we are surrounded by those events and facts which are there because of the physical matter of the earth. This matter is moulded and shaped in the most varied manner so as to be adapted to the beings of the kingdoms of nature, up to the human form itself. The essential element in all this is the physical matter of the earth. Today — because we shall immediately have to speak about its opposite — let us call this matter the physical substance of the earth, comprising all that provides the material basis for the various earthly forms; and then let us differentiate from it everything in the universe which is the opposite of this physical substance, namely *Cosmic* substance. This last is the basis not only of our own soul, but also of all those formations in the universe which, as spiritual formations, are connected with physical formations.

It is not right to speak only of physical matter or physical substance. Think only of the fact that we must place into the total picture of the world the beings of the higher hierarchies. These beings of the higher hierarchies have no earthly substance, no physical substance, in what in their case we would call their bodily

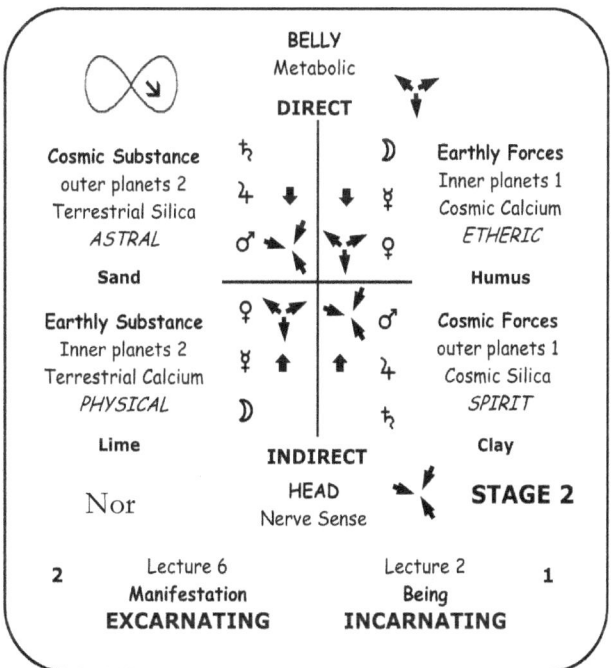

	BELLY Metabolic DIRECT	
Cosmic Substance outer planets 2 Terrestrial Silica ASTRAL	♄ 4 ☿ ♂ ♀	☽ Earthly Forces Inner planets 1 Cosmic Calcium ETHERIC
Sand		Humus
Earthly Substance Inner planets 2 Terrestrial Calcium PHYSICAL	♀ ☿ ☽	♂ Cosmic Forces outer planets 1 Cosmic Silica SPIRIT
Lime	♄ INDIRECT HEAD Nerve Sense	Clay
Nor		STAGE 2

2   Lecture 6   Lecture 2   1
         Manifestation   Being
         EXCARNATING   INCARNATING

123

nature. What they have is *Cosmic* substance. When we look upon what is earthly, we become aware of physical substance; when we can look upon what is outside the earthly, we become aware of *Cosmic* substance.

Today people know little of *Cosmic* substance. That is why they also speak of that earth-being, who belongs both to the physical and the spiritual — the human being — as though he, too, only possessed physical substance. This, however, is not the case. Man bears both *cosmic* and physical substance in himself in so remarkable a way as to astonish anyone who is not accustomed to pay heed to such matters. If, for example, we consider that element in man which leads him into movement, namely what is connected with the human limb-system and its continuation inwards as digestive activity, then it is incorrect to speak primarily of physical substance. You will soon understand this still more exactly. We only speak correctly about the human being when we regard the so-called lower part of his nature as having as its basis what is in fact *Cosmic* substance. So that, if we were to represent the human being schematically, we would have to say: The lower man actually shows us a formation in *cosmic* substance, and the more nearly we approach the human head, the more is man formed of *Earthly* substance. Basically the head is formed out of *Earthly* substance; but of the legs — grotesque though this may sound — it must be said that essentially they are formed of *cosmic* substance. So that, when we approach the head, we must represent the human being in such a way that we allow *cosmic* substance to pass over into *earthly* substance; in the human head where in particular *earthly* substance is contained. *Cosmic* substance, on the other hand, is diffused in a particularly beautiful way just where — if I may put it so — man stretches out his legs, stretches out his arms, into space. It is really as though the most important matter for arm and leg is precisely this being filled with *cosmic* substance, as if this is their essence. In the case of arm and leg it is really as though the physical substance were only swimming in the spiritual substance, whereas the head presents a compact formation composed of physical substance. In a form such as man possesses, however, we must differentiate not only the *substance,* but also *the forces.*

And here again we must distinguish between *cosmic* forces and earthly, physical forces.

In the case of the forces, things are completely reversed. Whereas for the limb-system and digestion the substance is *cosmic*, the forces in the limbs, for instance in the legs, are heavy, *earthly* forces. And whereas the substance of the head is physical, the forces active within it are *cosmic*. Spiritual forces play through the head; physical forces play through the *cosmic* substance of the limb and metabolic system in man. The human being can only be fully understood when we distinguish in him the upper region, his head and also the upper part of the breast, which are actually *earthly* substance worked through by *cosmic* forces (I must mention that the lowest spiritual forces are active in the breathing). And we must regard the lower part of man as a formation composed of *cosmic* substance, within which *earthly* forces are working. Only we must be clear as to how these things are interrelated in man, for the human being also projects his head-nature into his whole organism, so that the head — which is what it is because it is composed of *earthly* substance worked through by *cosmic* forces — the head also projects its entire nature into the lower part of the human being; and what man is because of his *cosmic* substance, in which *earthly* forces are at work, this, on the other hand, plays upwards into the upper part of the organism. In these activities in the human being there is mutual interaction. Man can in fact only be understood when he is regarded in this way, as composed of *earthly—cosmic* substantiality and *earthly – cosmic forces*, that is to say what is of the nature of forces.

This is something of great significance. For if we look away from external phenomena, and enter into the inner being, it becomes clear to us, for instance, that no irregularities can be allowed to enter into this apportioning of what is of the nature of substance and of forces in the human being.

Earthly Substance stays in the Metabolism

If, for example, what should be pure substance, pure *cosmic* substance in man, is too strongly penetrated by physical matter, by *earthly* substance — if, that is to say, *earthly* substance which should in fact tend upwards towards the head, as *'Ash'* , makes itself too strongly felt in the metabolism — then digestion becomes too strongly affected by the head-system, and man becomes ill; certain quite definite types of illness then arise. And then the task of healing consists in paralyzing, in driving out, the *earthly* substance-formation which is intruding into the *cosmic* substantiality. On the other hand, when man's digestive system, in its peculiar manner of being worked through by *earthly* forces in *cosmic* substance, when this digestive system is sent up towards the head, then the head becomes, as it were, too strongly spiritualised, then there sets in a too strong spiritualisation of the head, *causing migraines.* And now, because this also presents a condition of illness, care must be taken to send enough *ashed* physical forces of nourishment to the head, so that they reach the head in such a way that they do not become spiritualized.

Anyone who turns his attention to man in health and sickness will very soon be able to perceive the usefulness of this differentiation, if he is really concerned with truth, and not with external illusion. But something essentially different also plays into this matter.

Cosmic and Earthly in Nature (1)

In the immediate vicinity of the earth, we have the Moon and the other planets. The old instinctive science which reckoned the Sun, as one of the planets had one of the following sequence Moon, Mercury, Venus, Sun, Mars, Jupiter, Saturn. Now without going any further into the astronomical aspect of the subject, *other than to say this order is based upon the length of the cycle of the planet,* I wish to point to the relation which exists between planetary life and life on the earth. If we consider life on the earth in general the

♄	Saturn
♃	Jupiter
♂	Mars
♀	Venus
☿	Mercury
☾	Moon

first thing we have to take into account is the very important part played by the what I might call the life of the silicious substance in the world. You will find this silicious substance in the very beautiful mineral quartz enclosed in prismatic and pyramidal forms. Quartz is silicious substance combined with oxygen; remove the oxygen mentally, and you have the so called silicon. This silicon is regarded by modern chemistry as one of the elements (oxygen, etc,) and when united with oxygen may be regarded as a chemical substance. But we must not forget that this silicon which lives in the mineral quartz makes up from 47% to 48% of the crust of the Earth, i.e. a higher percentage than that of any other substance on earth, oxygen, for example, amounting only to 27% to 28%. Now silicon, in the form in which it appears in such stony substances as quartz, does not at first seem to possess very much importance if we consider only the material of the soil of earth with its plant growth. Quartz is not soluble in water - the water trickles through it. It thus seems to have no connection with the ordinary commonplace view of "conditions of life". But if you take the Equisetum (horsetail) you will find that it consists of 90% of silicon (the same substance of which quartz consists) in very fine distribution through its form. This shows the enormous importance which this substance, silicon must have. It forms nearly one half of everything on the earth, And yet so completely has its importance been overlooked that its use has been neglected even where it can have the most beneficent results. Silicon forms an essential constituent of many remedies used in Anthroposophical therapy. A whole series of diseases are treated either internally or by baths with this substance, the reason being that what appears in the form of abnormal conditions of the sense organs, (it only appears there, it does not really lie there) the internal sense organs, as a cause of pain is strangely accessible to the influence of silicon. And in general silicon plays the greatest conceivable part in what has been called by the old-fashioned name of the "household of Nature". For it is present not only in quartz and other stones, but in a highly refined state in the atmosphere. Indeed it is present everywhere. One half of the earth at our disposal consists of silicon, what then is the function of this substance?

To answer this question let us assume that our earth contained only half of the quantity of silicon which it actually does possess — *giving a Calcium dominant environment.* We should then have plants in more or less pyramidal form: the blooms would be atrophied and indeed all plants would assume generally the shape of the cacti which strikes us as so abnormal. The cereals would look grotesque; their stems would grow thick and fleshy towards the base, but the ears would be emaciated and without grain.

So much for silicon. On the other hand in every part of the earth, although not in such abundance as is silicon, we find lime and their allied substances (limestone, Potash and Sodium). If these were present in small proportions — *and thus have a Silica dominant environment* — we should have plants whose stems were only narrow and twisted we should have only creepers. There would be blooms of course but they would be useless and yield nothing of any food value.

It is only through the balance of these two formative *processes* - as embodied in these two substances, silicon and limestone - that plant life can flourish in the form in which we know it today. *This working together however must be identified in two ways. The first in how the external processes , as World energetic bodies work ONTO life, (discussed elsewhere) while the second is when the same processes function WITHIN the Physical bodies of lifeforms.*

Now, everything silicious contains *Cosmic* forces that come not from the earth but from the so-called distant planets- Mars, Jupiter and Saturn - the planets beyond the Sun. These planets work indirectly upon plant-life through silicon and allied substances. But the planets near the Earth- Moon, Mercury and Venus, send out *Earthly* forces into the plant life and animal life on earth through the medium of the limestone and kindred substances. Thus of any cultivated field it may be said that the **forces** of both silicon and limestone are at work in it. The silicon mediates the *Cosmic Force* influences of Mars, Jupiter and Saturn, *from below* and the limestone those *Earthly Forces* of Moon, Venus and Mercury, *from above*.

Now let us turn to the plants themselves. There are two things to notice about all plants. The first is that the plant world as a whole and every single species have the power to perpetuate their kind and develop the force of reproduction, etc. The second is that the plant as a member of a relatively low order of nature serves as nourishment for members of higher orders. These two fundamental tendencies seem at first to have little to do with one another. For if we only look at the passing on of the step from parent plant to offspring and so on, it is a matter of indifference to the formative forces of Nature whether or not the plant is used for food. The two interests (i.e. of nature and Man) are completely different, and yet the forces of Nature act in such a way that the inherent powers of reproduction and growth and of producing generation after generation of plants, are active in the

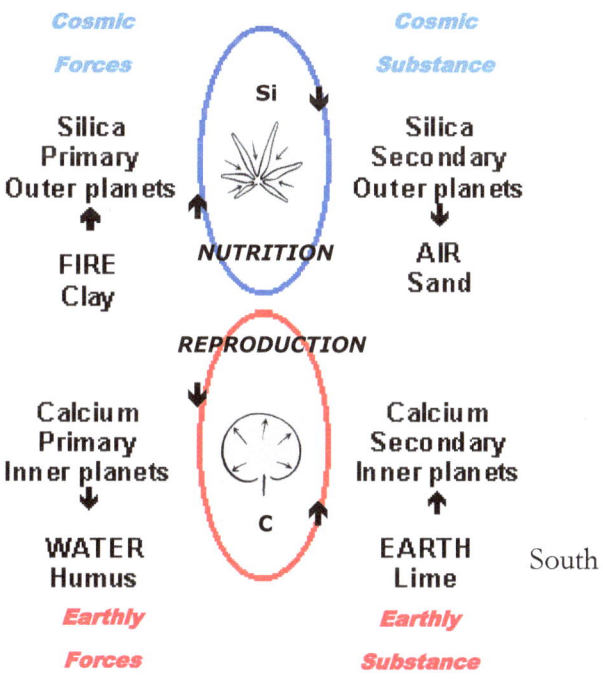

Cosmic Forces

Cosmic Substance

Si

Silica
Primary
Outer planets

Silica
Secondary
Outer planets

FIRE
Clay

NUTRITION

AIR
Sand

REPRODUCTION

Calcium
Primary
Inner planets

Calcium
Secondary
Inner planets

C

WATER
Humus

EARTH
Lime

South

Earthly Forces

Earthly Substance

influences exercised upon earth by the Moon, Venus and Mercury, through the mediation of limestone. If we consider plants which are not used for food, which do nothing but reproduce themselves, we take *special* interest in the forces of Venus, Mercury and Moon, related to reproduction. But in the case of plants which are eminently suitable for food, because their substances have become perfected to the point of forming foodstuffs, for human and animal consumption, it is the planets Mars, Jupiter and Saturn that are working through the medium of silicon. Silicon opens up the being of plant to the expanses of the Universe, it awakens the plant's senses, so that it absorbs the formative forces bestowed by the distant planets, Mars, Jupiter and Saturn. From the sphere of Moon, Venus and Mercury on the other hand, the plant absorbs only that which makes it capable of reproducing itself. Now this seems at first to be just an interesting theory. But every insight taken from a wider horizon leads us quite naturally from theory to practice.

For instance in what way can the activities of Moon or Saturn be modified in their influence on plants? If we observe the course of the year, we shall find that on some days there is rain and on others none. All that the modern physicist observes is the fact that on rainy days more water falls on the Earth than on dry days! Water moreover is to him something abstract consisting of oxygen, hydrogen, and nothing more. If water is decomposed by electrolysis it is split into two substances, each of which acts in its own way. But this tells us nothing about water. There is much more hidden in water than appears in the chemical properties of hydrogen and oxygen. Water by its very nature is eminently fitted to bear along with it the forces coming from the Moon on to the Earth. So It comes about that it is water which distributes the lunar forces throughout the earthly realm. There is a certain kind of relation between the Moon and the water on the Earth. Let us suppose that after a rainy spell there is a full moon. Now the forces coming from the Moon when it is full causes something tremendous to happen on Earth. They shoot right into the whole growing forces of the vegetable kingdom. They cannot do so if there has not been a rainy spell beforehand. We must always realise the importance of sowing seed after rainy days followed by the full moon and we should never work at random (true, something will always come up). The question: How to connect our seed-sowing with rain and full Moon has definite practical importance, because the forces that come from the Full Moon work powerfully and abundantly on certain plants after rain but only weakly and sparingly after a spell of sunny weather. The old adages of husbandry contained such knowledge.

Furthermore around the Earth we find the atmosphere. In addition to consisting of air, the atmosphere has the property of being sometimes warm and sometimes cold. At times there is certainly accumulation of heat which, if the tension becomes too great, may discharge itself in a thunderstorm. Now what can we say about warmth? Spiritual observation shows that while water has no relation to silicon, warmth is so powerfully related to it that it enhances the activity of the forces working through silicon, namely , the forces coming from Saturn, Jupiter and Mars. These forces coming from Saturn, Jupiter and Mars have to be valued on quite a different scale from that adopted in the case of Moon Venus and Mercury, for it must be remembered that Saturn takes thirty years to go round the Sun, while the Moon takes only about twenty-eight days to pass through all its phases. Thus Saturn is only visible for fifteen years, consequently stands in quite another relation to the growth of plants compared with the Moon. As a matter of fact Saturn is not only active when it is shining down on the Earth, it is also active when its rays have to pass from below, as it were, through the Earth.

The strength with which the Saturn forces influence plant life on Earth always depends upon the warmth-condition of the air. If the air is cold they cannot reach the plants, if the air is warm they can. How then can we see their influence at work in the plant? We see it not in the annuals but in the perennials; not in those plants which grow up and die in the course of one year leaving only their seed behind them but in those which are perennial. It is the latter whose growth Saturn promotes with the help of the warmth forces of the Earth. The effect of these forces working through the mediation of warmth, is to be seen, for instance in the bark or cortex of trees and in everything that makes the plant a perennial. When the lives of plants are limited to the short span of a single year, it is because of the relation in which those plants stand to the planets with short periods of revolution. On the other hand, that which emancipates itself from the fleeting process and is made permanent in the formation of bark around the growing trees is connected with the planetary forces working through the mediation of warmth and cold, and the periods of revolution in these cases are long. Thirty years in the case of Saturn, twelve in the case of Jupiter. Again it is well for anyone who wants to plant an oak tree to know something of the periodicity of Mars, for an oak tree planted during the appropriate period of Mars will thrive much better than one planted unthinkingly, at any moment that happens to be convenient. Or, if you have a plantation of conifera, where the Saturn forces play so great a part, it will make all the difference if the trees are planted when Saturn is in the so-called ascending period or at another time. Anyone who has insight into these matters can tell quite accurately in the case of plants that are doing well or badly whether or not they have been tended with a right understanding of their relation to planetary forces. For what is not always obvious to the external eye is revealed to more intimate observation.

I will indicate the surface of the Earth diagramatically by this line. The surface of the Earth is generally regarded as mere mineral matter — including some organic elements, at most, inasmuch as there is formation of humus, or manure is added. In reality, however, the earthly soil as such not only contains a certain life — a vegetative nature of its own — but an effective astral principle as well; a fact which is not only not taken into account to-day but is not even admitted nowadays.

But we can go still further. We must observe that this inner life of the earthly soil (I am speaking of fine and intimate effects) is different

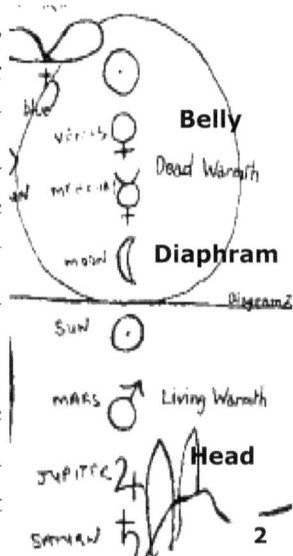

Belly

Dead Warmth

Diaphram

Sun

Living Warmth

Head

132

in summer and in winter. Here we are coming to a realm of knowledge, immensely significant for practical life, which is not even thought of in our time.

Taking our start from a study of the earthly soil, we must indeed observe that the surface of the Earth is a kind of organ in that organism which reveals itself throughout the growth of Nature. The Earth's surface is a real organ, which — if you will — you may compare to the human diaphragm. (Though it is not quite exact, it will suffice us for purposes of illustration). We gain a right idea of these facts if we say to ourselves: Above the human diaphragm there are certain organs — notably the head and the processes of breathing and circulation which work up into the head. Beneath it there are other organs.

If from this point of view we now compare the Earth's surface with the human diaphragm, then we must say: In the individuality with which we are here concerned, the head is beneath the surface of the Earth, while we, with all the animals, are living in the creature's belly!

Whatever is above the Earth, belongs in truth to the intestines of the "agricultural individuality," if we may coin the phrase. We, in our farm, are going about in the belly of the farm, and the plants themselves grow upward in the belly of the farm. Indeed, we have to do with an individuality standing on its head. We only regard it rightly if we imagine it, compared to man, as standing on its head. With respect to the animal, as we shall presently see, it is a little different.

Why do I say that the agricultural individuality is standing on its head? For the following reason. Take everything there is in the immediate neighbourhood of the Earth by way of air and water vapours and even warmth. Consider, once more, all that element in the neighbourhood of the Earth in which we ourselves are living and breathing and from which the plants, along with us, receive their outer warmth and air, and even water. All this actually corresponds to that which would represent, in man, the abdominal organs. On the other hand, that which takes place in the interior of the Earth beneath the Earth's surface — works upon plant

-growth in the same way in which our head works upon the rest of our organism, notably in childhood, but also throughout our life. There is a constant and living mutual interplay of the above-the-Earth and the below -the-Earth.

And now, to localise these influences, *into the Physical Formative Forces*, I beg you to observe the following. The *Force* activities above the Earth are immediately dependent on Moon, Mercury and Venus supplementing and modifying the influences of the Sun. The so-called "planets near the Earth" extend their *force* influences to all that is above the Earth's surface. On the other hand, the distant planets — those that revolve outside the circuit of the Sun — work upon *the force activities* that *are* beneath the Earth's surface, assisting those influences which the Sun exercises from below the Earth. Thus, so far *as the Force aspects* of plant-growth is concerned, we must look for the influences of the distant Heavens beneath, and of the Earth's immediate cosmic environment above the Earth's surface.

Once more: all that works inward from the far spaces of the Cosmos, *as Cosmic Forces*, to influence the growth of plants, works not directly — not by direct radiation — but in this way: It is first received by the Earth *from the Cosmic Substance*, and the Earth then rays it upward again. Thus, *the Cosmic Force* influences that rise upward from the earthly soil — beneficial or harmful for the growth of plants — are in reality cosmic influences rayed back again. *The Cosmic Substance is* working directly in the air and water over the Earth. The direct radiation from the Cosmos, *coming via the Cosmic Substance* is stored

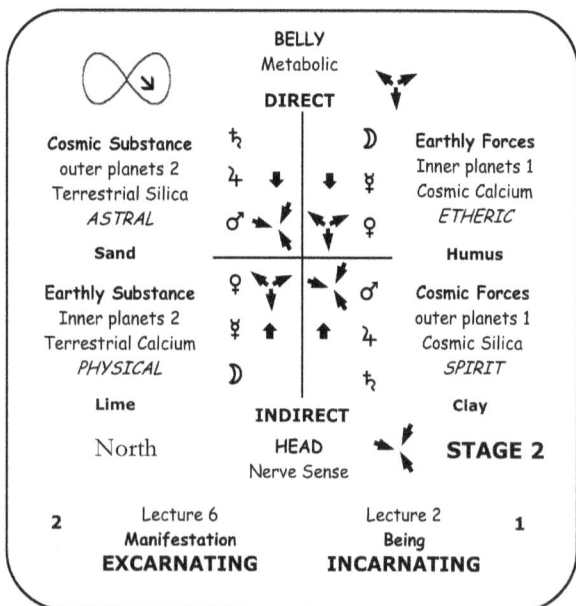

Diagram contents:

BELLY / Metabolic / DIRECT

Cosmic Substance / outer planets 2 / Terrestrial Silica / ASTRAL / Sand — ♄ ♃ ♂

Earthly Forces / Inner planets 1 / Cosmic Calcium / ETHERIC / Humus — ☽ ☿ ♀

Earthly Substance / Inner planets 2 / Terrestrial Calcium / PHYSICAL / Lime — ♀ ☿ ☽

Cosmic Forces / outer planets 1 / Cosmic Silica / SPIRIT / Clay — ♂ ♃ ♄

North

INDIRECT / HEAD / Nerve Sense

STAGE 2

2 — Lecture 6 / Manifestation / EXCARNATING

Lecture 2 / Being / INCARNATING — 1

up beneath the Earth's surface *as Cosmic Forces* and works back from there. Now these relationships determine how the earthly soil, according to its constitution, works upon the growth of plants. (We shall take plant-growth to begin with, and afterwards extend it to the animals).

With the Substance stream we must learn to distinguish those *Earthly* Forces which arise in the cosmos but are absorbed by the earth and work upon plant-growth as *Earthly Substance,* from within the earth. These forces come from Mercury, Venus and Moon and act not directly, but through the mediation of the earth. They must be taken into account if we wish to follow up how the mother- plant gives rise to a daughter- plant, and so on. On the other hand, we have to consider the *Cosmic Substance* forces taken by the plant from the outer-earthly, and brought to it by way of the atmosphere from the outer planets. Broadly speaking, we may say that the forces coming from the nearer planets are very much influenced by the workings of lime in the soil, while those coming from the distant planets fall under the influence of silicon. And, in fact, *the Cosmic Force* workings of silicon, even though they proceed from the earth, act as mediators of the forces coming from Jupiter, Mars and Saturn, but not for those of Moon, Mercury and Venus, *which also work in the Earth, as Earthly Substance.*

Now I want you to imagine that Diagram No. 9 represents the earth level, where the influences of Venus, Mercury and Moon; enter as *Earthly Forces* into the earth and stream again from below upwards as Earthly Substance. These are the forces which cause the plant to grow during the season, later produce the seed, and by means of this seed a new plant', a second plant, then yet a third and so on. (I indicate this schematically). All this goes into the power of reproduction and streams on into the succeeding generations. The

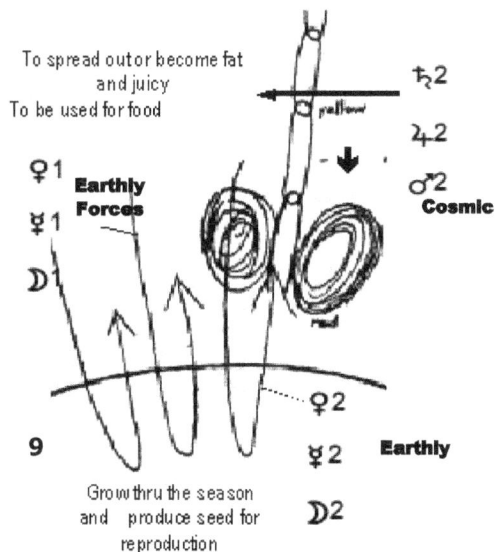

To spread out or become fat and juicy
To be used for food

♀1
☿1 **Earthly Forces**
☽1

9

Grow thru the season and produce seed for reproduction

yellow
red

♄2
♃2
♂2 **Cosmic**

♀2
☿2 **Earthly**
☽2

Cosmic Substance forces, however, which take the other path, remaining above the earth level, come from the *secondary* distant planets. I can draw this schematically in this way. These forces cause the plant either to spread into its surroundings or to become fat and juicy, to build matter into itself such as we can use for food because it is produced again and again in a continuous stream.

Take for example the flesh of fruit - an apple or a plum - which we can break off and eat; all this is due to the *secondary* workings from the distant planets.

Consider the earthly soil. To begin with, we have those *Cosmic Substance* influences that depend on the farthest distances of the Cosmos — the farthest that come into account for earthly processes. These effects are found in what is commonly called sand and rock and stone. Sand and rock — substances impermeable to water, which, in the common phrase, "contain no foodstuffs" — are in reality no less important than any other factors. They are most important for the unfolding of the growth-processes, and they depend throughout on the influences of the most distant cosmic *substance* forces. And above all — improbable as it appears at first sight — it is through the sand, with its silicious content, that there comes into the Earth what we may call the life-ethereal and the chemically influential elements of the soil. These influences then take effect as they ray upward again from the Earth, *as Cosmic Forces.*

The way the soil itself grows inwardly alive and develops its own chemical processes, depends above all on the composition of the sandy portion of the soil. What the plant-roots experience in the soil depends in no small measure on the extent to which the cosmic life and cosmic chemistry are seized and held by means of the stones and the rock, which may well be at a considerable depth beneath the surface. Therefore, wherever we are studying plant growth, we should be clear in the first place as to the geological foundation out of which it arises. For those plants in which the root-nature as such is important, we should never forget that a silicious ground — even if it be only present in the depths below — is indispensable. I would say, thanks be to God that silica is very widespread on the Earth — in the form of silicic acid, for instance, and in other compounds. It constitutes 47-48% of the surface of the Earth, and for the quantities we need we can reckon practically everywhere on the presence of the silicic activity.

But that is not all. All that is thus connected, by way of silicon, with the root-nature, must also be able to be led upward through the plant. It must flow upward. There must be constant interaction between what is drawn in from the Cosmos by the silicon, as *Cosmic Forces* and what takes place — forgive me! —in the "belly" up above; as *Cosmic Substance* for by the latter process the "head" beneath must be supplied with what it needs. The "head" is supplied *with Cosmic Forces* out of the Cosmos, but it must also be in mutual interaction with what is going on *with the Cosmic Substance* in the "belly," above the Earth's surface. In a word, that which pours down from the Cosmos and is caught up beneath the surface, as *Cosmic Forces,* must be able to pour upward again. And for this purpose is the clayey substance in the soil. Everything in the nature of clay is in reality a means of transport, for the influences of cosmic *force* entities within the soil, to carry them upward again from below.

What comes from beneath as good or bad vegetable growth are really the cosmic *force* influences which are reflected from below; whereas in the air and water above the earth, the cosmos exercises its power directly. The direct cosmic in-streaming, *Cosmic Substance,* is stored up beneath the earths surface, and from there works back, *as Cosmic Forces.* The inherent qualities of the soil affecting the growth of plants are dependent upon these stored up *cosmic force* influences. The soil still retains in it the effects of influences dependent upon the most remote parts of the Cosmos, which need to be considered in connection with the Earth.

Let us suppose that we want to **hold back these *Cosmic* forces**, which work upwards from the root through the stem into the leaves, and store them up in the region of the root. This possibility is no longer fully open to us in the present epoch of our earth, since genera and species of plants have been so firmly established. Formerly, in ancient epochs when men could easily transform one plant into another, this possibility had to came greatly into consideration. Today we consider it only from the point of view of finding out the condition favourable to a given plant. How can we then set about preventing these forces from pushing upwards into blossom and fruit? How can we in addition hold back the development of stem and leaf within the formation of the root? We must place such a plant on sandy soil. For silicon or flint holds back the Cosmic Forces and even gathers them. Now the potato plant is one in which the growth of leaf and stem is held back. The potato is a root-stock. The forces that form leaf and stem are held fast in the potato itself. The potato is not a root but a stem which has been held back. Potatoes must therefore be planted on sandy soil; this is the only way of holding back the Cosmic Forces in them.

Let us plant two experimental beds with wheat and sainfoin respectively. Then, if silica *sand* has been added to the soil, you will be able to observe that the wheat (a plant whose natural and permanent tendency it is to

Sand - holding back the upward Silicia
A Homeopathic that includes Sand sprayed on Celery

produce seed) is being hampered in its seed formation. In the case of the sainfoin you will also see that the seed formation is either completely suppressed or is retarded. In such "experiments" you can always take the effects on the cereal as the basis for comparison with the corresponding effects on sainfoin as representing leguminous plants. In this way very interesting experiments can be made in seed-formation.

QUESTION: Does it make any difference whether the soil underneath is sand or clay? Often people put a ground layer of clay where the manure is to be, so as to make the ground impervious.

ANSWER: It is quite true that different kinds of soil have a definite influence which proceeds from the particular qualities of the soil in question. A sandy soil does not retain water; it is therefore necessary to put some clay with it before laying the manure on it. If, on the other hand, you have a clay soil, you should break it up and strew sand over it. A middle course would be to have alternate layers of sand and clay. Then you have the earth consistency as well as the "watery influences. Without this combination of the two kinds of soil the water will percolate away. For the same reason, loose soil should certainly not be used as a foundation for the manure heap as it would have no value for the manure placed over it; in this case it is better to make your own foundation.

However this cosmic *force* upward flow is not enough by itself. There must also be present the opposite, which I could call the Earthly or terrestrial *Forces* streaming downwards. All that undergoes a kind of external digestion in the "belly" (the processes above the surface throughout Summer and winter are indeed a kind of digestion in relation in the growth of plants!) has to be drawn down into the earth. All *Earthly* forces produced by the action of water and air above the Earth, and also the substances in delicate homoeopathic distribution called from there, are drawn down into the earth by lime present in it in greater or smaller proportions. The lime content of the soil and the distribution of lime in homoeopathic dilution above the surface - these are the factors which have the task of leading the terrestrial forces down into the soil.

These things will take on a very different aspect in future when we shall have a real science concerning them and not only the scientific guesswork of today: it will be possible then to give exact information. We shall then know that there is a great, an immense difference between the warmth that exists above the surface of the Earth and which stands within the sphere of the *primary* influence of the Sun, Venus, Mercury and Moon and the warmth which makes itself felt within earth and which stands under the primary influence of Mars, Jupiter and Saturn. These two kinds of warmth which we may call the "blossom and leaf warmth" and the "root-warmth" respectively, are completely different from one another - so much so, indeed, that we can describe the warmth above the Earth as a "dead" warmth, the warmth below the Earths surface a "living" warmth. The warmth below the surface, especially during Winter contains an inner vital principle. If we human beings had to experience in ourselves this living warmth, which works within the soil we should all become immensely stupid because in order that we may be intelligent beings, dead warmth has to be supplied to our bodies. But at the moment when the limestone and other substances enable warmth to be drawn into the soil and to change from outer into inner warmth it passes over into a condition of gentle aliveness. It is recognised to-day that there is a difference between the air which is above the Earth and that which is below the surface, but the difference between warmth above the Earth and that below the surface has been overlooked. It is generally known that the air under the Earth contains more carbonic acid, while that above the Earth contains more oxygen; but the reason for this is not known. It is that the air, as it is drawn into the earth, is penetrated by a gentle aliveness. This is true both of warmth and of air. They both receive a tiny spark of life as they pass into the earth. It is different in the case of water and of the solid earth element itself. Both of these have less life inside the Earth than they have when above its surface. They become "more dead", they lose something of their life they had outside. But it is precisely this circumstance which exposes them to the influences of the most distant cosmic forces.

Now with regard to the cultivation of the soil there is a point of great importance which must be thoroughly understood. It is a point I have often dealt with amongst Anthroposophists. It is that we know the conditions which the forces of the cosmic spaces can work upon the earthly realm. Let us begin with seed formation. The seed which gives rise to the embryo of the plant is generally regarded as a molecular structure of exceptional complexity, and science lays great stress upon this interpretation. The molecules it is said have a certain structure, in simple molecules it is simple, in complicated molecules it becomes more and more complex, until we come to the extreme complexity of the albuminous or protein molecule. People stand in wonder and astonishment at the enormous complexity of the structure supposed to exist in the seed.

They do so because they reason as follows. The albumen (or protein) molecule, they say, must be of enormous complexity, for the organism in succeeding plants arises from it. This organism is enormously complex, and since its structure was determined by the embryonic conditions of the seed, the latter's microscopic or ultra-microscopic content must also have a structure of enormous complexity. Well, it is complex indeed in the beginning. As the earthly albumen is formed, its molecular structure is driven to the utmost complexity; but this alone would never give rise to a new organism. For the organism arising from the seed does not proceed by a mere continuation in the off- spring of what was present in the parent plant or animal.

What happens is that when the embryonic structure has reached the highest stage of complexity in the earth domain, *during pollination*, it falls to pieces and becomes a "little chaos". It breaks up and dissolves, one might say, into "world-dust". And when this little chaos of world-dust is there, the whole surrounding cosmos begins to work upon it to stamp it with its own image and to build up in it a structure conditioned by the forces of the Universe working in upon it from every side (see drawing no. 3). Thus the seed becomes an image of the Cosmos. Every time this happens, and seed formation is carried through to the point of chaos, the new organism is built up from the seed-chaos by the activity of the cosmos. The parent organism has only the tendency to bring the seed to such cosmic position that through its affinity with this cosmic position the *appropriate* forces will act in the proper direction so that, eg a dandelion will give rise to another dandelion and not a berberis.

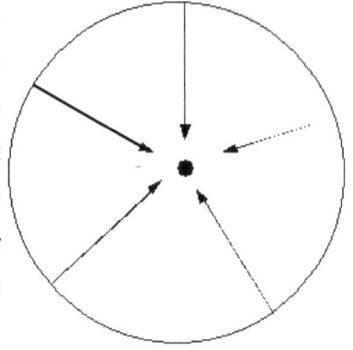

But the new thing that is built up is always the image of some cosmic constellation. It is built up out of the cosmos. And if in the Earth we would make effective the forces of the cosmos, we must drive the earthly elements into the state of greatest possible chaos. This has to be the case whenever we want the cosmos to act upon our Earth. In the case of plant -growth this is in a certain sense provided for by nature herself. But just because every new organism is built up by the Cosmos it is necessary that **the cosmic principles must be allowed freedom to work in the organisms until the seed-formation is completed.**

If for example, we plant the seed of a given plant in the earth the seed contains the impress of the whole cosmos from a particular cosmic direction , which means that it came under the influence of a particular *star* constellation and received its particular *archetypal species* form. At the moment when the seed is placed in the soil it is strongly worked upon by the *Earthly Forces* and it is filled with the longing to deny the Cosmic Forces, in order that it may spread and grow in all directions. For the *Earthly* Forces above the surface of the Earth, do not want the plant to retain this cosmic form. The seed had to be driven to the point of chaos *at pollination;* but now that the plant is sprouting it is necessary to oppose the *Earthly* to the Cosmic Forces, which live as the form of the plant inside the seed. For the Cosmic Forces must be opposed and balanced, as it were, by the *Earthly* Forces. We must help the plant to become more akin to the Earth in its growth. This can only be done by introducing into the plant some form of living earthly matter which has not yet reached the state of chaos and seed formation, life which has been held up in a plant before the seeds have been formed. For this purpose a rich humus formation comes to man's assistance In those districts that are fortunate enough to possess it. Man can hardly find any artificial substitute for the fertility given to the soil by Nature through **humus**. What causes formation of humus? It arises from the absorption of remnants of living plants into the whole process of Nature. These remnants have not yet reached the state of chaos and respect the Cosmic Forces, as it were. If humus is used for the growth of plants the *Earthly* Forces are held fast within them. The Cosmic Forces then work only in the upward stream that terminates in seed-formation. While the *Earthly* Forces work in the development of flowers, leaf and so on, the cosmos only radiates its influence into all this.

Let us suppose that we have before us a plant growing out of its own root. At the top end of the stem comes the grain of seed, while the leaves and blossoms spread out sideways. Now, in the leaf and the blossom *the above ground activities* are working in giving shape *through the Cosmic Substance* and filling it with matter, *via the Earthly Forces* ; the reason why a leaf grows or a grain swells and takes up the substance inside it is to be found in the Earthly Forces which we lead to the plant and which have not yet reached the point of chaos. The seed, however whose *Cosmic* forces work upwards through the stem - vertically (not rotating around it , *which come from the Earthly Substance and shows as the tissue formation of leaves*) - radiate the cosmic forces into leaves and blossoms. One can actually see this. We have only to look at the green leaves of a plant. In their shape *we see the Cosmic Substance,* in the substances filling them *we see the Earthly Forces* and in their green colour, the leaves bear *the above ground elements*. But they would not be green, if they had not within them the cosmic force of the Sun. And now look at the coloured blossoms. In these the cosmic forces of the Sun are not working alone, but are supported by the primary distant planets Mars *(1),* Jupiter *(1)* and Saturn *(1).* If we regard the growth and development of plants from this point of view, we shall see the redness of the rose as the force of Mars, the yellow of the sunflower (so-called only because of its shape) as the *cosmic* force of Jupiter. It should be called the Jupiter flower, for it is the force of

Jupiter that reinforces the solar force and brings forth the white and the yellow colours in the flowers. The blue of chickweed or chicory flower is the effect of Saturn reinforcing the effect of the Sun. Thus we can see Mars in the red coloured flower, Jupiter in the yellow, Saturn in the blue, while in the green colour of the leaf we see the Sun Itself.

QUESTION: Should we take any special measures to strengthen the tendency of the seed to be "driven into chaos"?

ANSWER: One can strengthen it but there is no need to do so, because if seed formation comes about at all then there is always a maximum of "chaos". It therefore does not need to be strengthened. Any necessary strengthening must be done to the manure; but it is not necessary for the seed formation. We could, of course, do something by making the soil more silicious, *with clay and sand*. For it is through silica *clay* that the Cosmic Forces work which have been absorbed into the earth. -"One could do it in this way, but I do not think that it is necessary, *unless seed formation does not occur, which can be the case with an overly sandy soil.*

Roots

But the same powers which appear as colour in the flower are also at work especially strong in the root. Here once more the forces living in the distant planets are active within the soil. If we pull a plant out of the ground we may see that in the *tap* roots there is Cosmic Force, in the blossom mostly the *earthly force* element and only in the finest shading by the colour can the cosmic *substance* element be seen. The *Earthly* Forces on the other hand, if *working* actively in the root, cause the root to push out into *a ramified* form. For the form of the plant is determined by factors arising in the realm of earth. It is the *Earthly* Forces that causes the form to spread. When the root develops and divides, it is due to the *earthly* forces working downwards just as the cosmic forces (in the case of the colour) work upwards. Single roots are therefore cosmic roots, whereas forked roots are due to the *earthly* forces working down into the soil, just as in colour the cosmic forces work upwards into the flowers, and the cosmic force of the Sun stands between the two. The Sun force works principally in the green leaves, in the interaction between blossom, root and in all that is between the two. Thus the Sun element really belongs to what we have called the diaphragm provided by the surface of the earth: whereas the Cosmic *Forces* belongs to the interior of the earth and works its way up into the upper part of the plant. The *Earthly* Forces above the earth, works downwards and is drawn into the plant with the help of the limestone. Plants which draw down the *Earthly Forces* into their roots through the lime, are those whose roots divide in all directions such as all herbs used for fodder, (but not turnips) and such as the sainfoin. Thus it should be possible, looking at the form of a plant and the colour of the flowers, to tell how much cosmic forces and how much *Earthly* Forces are at work in it.

Now let us assume that we find some means of **holding back the cosmic forces** within the plant. *I referred earlier to the role sand plays in this regard. However the plant kingdom also offers us another opportunity.* These *cosmic* forces will then be prevented from manifesting *themselves* by pushing up into flowers, but will live out their life in the region of the stem of the plant. Now wherein do these Cosmic Forces reside in the plant? They reside in the silicon. Take the **Equisetum.** It has this very property of attracting silicon and permeating itself with it. It is 90% silicon. Thus in this plant the cosmic element is present to a tremendous extent. It does not manifest itself in flowers, but in the growth of the lower part of the plant.

The ABC of everything concerning the growth of the plant consist therefore, in knowing what in any particular plant is cosmic origin and what is due to terrestrial forces. How can we make a soil more inclined to condense, as it were, the Cosmic Forces to retain them in root and leaf? How can we thin them out so that they can be sucked upwards into the blossoms and colour and even into the fruit and permeate them with a delicate taste? For the delicate taste in an apricot or plum is like the colour of a flower, both being due to the Cosmic Forces which have worked their way upward through the plant, *being met by the Cosmic*

Substance process above. In the apple you are literally eating Jupiter, in the plum you are eating Saturn.

It is imperative that our knowledge should penetrate to the actual structure of Nature. For example, man knows more or less what happens to air inside the earth, but he hardly knows any thing of what happens to light inside the Earth. He does not know that Silicon, the cosmic mineral, takes up light into the Earth and there makes it active, whereas humus, the substance closely allied to terrestrial life does not take up light and make it active in the earth but produces a lightless activity there. But these are the things which will have to become understood and known.

From this we are able to see how we must proceed if we are to influence plant-growth in one way or another. We have to take account of these two sets of forces.

Fungus - Rots

Now, from everything I have said on this subject, you will have gathered that the soil immediately surrounding a plant has a definite life of its own. These life forces are there and with them all kinds of forces of growth and tender forces of propagation not strong enough to produce the plant form itself, but still waiting with a certain intensity; and in addition all the forces working in the soil under the influence of the Moon and mediated through water. Thus certain important connections emerge, in the first place you have the earth, the earth saturated with water. Then you have the moon. The moon beams, as they stream into the earth, awaken it to a certain degree of life, they arouse "waves" and weavings in the earth's etheric element. The moon can do this more easily when the earth is permeated with water, less easily when the earth is dry. Thus the water acts only as a mediator. What has to be quickened is the Earth itself, the solid mineral element. Water, too, is something mineral. There is no sharp boundary, of course. In any case, we must have lunar influences at work in the earth. Now these lunar influences can become too strong. Indeed this may happen in a very simple

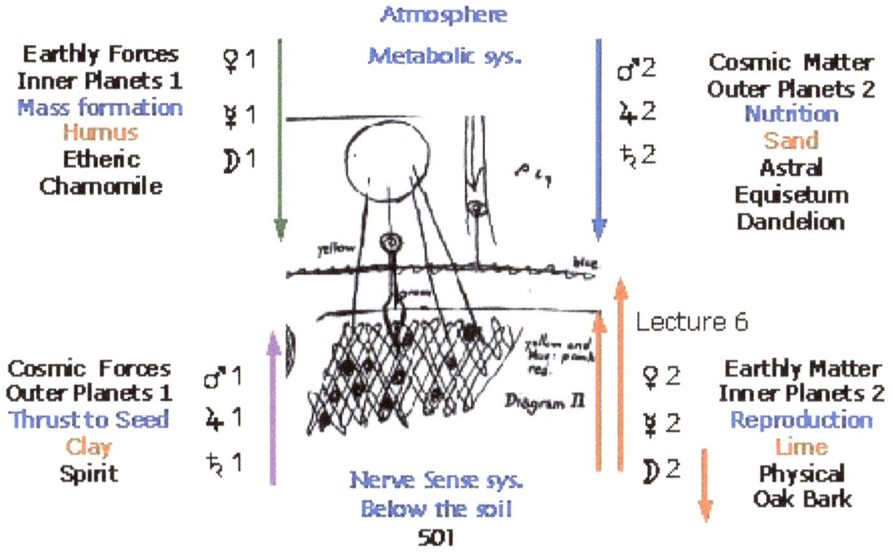

Atmosphere

Metabolic sys.

Earthly Forces	♀1		♂2	Cosmic Matter
Inner Planets 1			♃2	Outer Planets 2
Mass formation	☿1			Nutrition
Humus			♄2	Sand
Etheric	☽1			Astral
Chamomile				Equisetum
				Dandelion

yellow   blue

Lecture 6

Cosmic Forces	♂1		♀2	Earthly Matter
Outer Planets 1				Inner Planets 2
Thrust to Seed	♃1		☿2	Reproduction
Clay				Lime
Spirit	♄1		☽2	Physical
				Oak Bark

Diagram II

Nerve Sense sys.
Below the soil

501

manner. Consider what happens, when a very wet spring follows upon a very wet winter. The lunar force enters too strongly into the earth, which thus becomes too much alive. I will indicate this by red dot's. (See Diagram No. 11). Thus if the red dots were not here, i.e. if the earth were not too strongly vitalised by the moon, the plants growing upon it would follow the normal development from seed to 'fruit; there would be just the right amount of *reproductive* lunar force, *supported Earthly Substance,* distributed in the earth to work upwards, *upon the Cosmic Forces 'train' to* produce the requisite fruit seed. But let us suppose that the lunar influence is too strong - that the earth is too powerfully vitalised - then the *Earthly Substance* forces working upwards become too strong, and what should happen in the seed formation, *due to the Cosmic Forces activity,* occurs earlier. Through the very intensity the *Earthly Substance , the Cosmic* forces do not proceed far enough to reach the higher parts of the plant, but become active earlier and at a lower level. The lunar influence has the result that there is not sufficient *Cosmic Force* strength for seed formation. The seed receives a certain portion of the decaying *reproductive* life *force,* and this decaying life forms another level above the soil level. This new level is not soil, but the same *Earthly Substance* influences are at work there. The result is that the seed of the plant, the upper part of the plant

147

becomes a kind of soil for other organisms; parasites and fungoid formations appear in it. It is in this way that blights and similar ills make their appearance in the plant. It is through a too strong working of the moon that the *Cosmic* forces *'train'* working upward from the earth are prevented from reaching their proper height. The powers of fertilisation and *reproduction* depend entirely upon a normal amount of lunar influence. It is a curious fact that abnormal developments should be caused not by a weakening but by an increase of lunar forces. Speculation might well lead to the opposite conclusion. Looking at it in the right way shows that the matter is as I have presented it. What, then, have we to do? We have to relieve the earth of the excess of lunar forces in it. It is possible to relieve the earth in this way. We shall have to discover something which will rob the water of its power as a mediator and restore to the earth more of its earthiness, so that it does not take up an excess of lunar forces from the water. This is done by making fairly concentrated brew (or tea) of equisetum arvense (horse-tail), diluting it and using it as a liquid manure on the fields for the purpose of fighting blight and similar plant diseases. Here again only small quantities are required; a homeopathic dose is generally sufficient. As you will have realised, this is precisely where one sees how one department of life affects another. If, without indulging in undue speculation, we realise the noteworthy effects produced by equisetum arvense upon the human organism by affecting the function of the kidneys, we shall have, as it were, a standard by which to estimate what this plant can achieve when it has been transformed into liquid manure, and we shall realise how extensive its effects may be when even quite a small quantity is sprinkled about without the help of any special instrument. We shall realise that equisetum is a first-rate remedy. Not literally a remedy, since plants cannot really be ill. It is not so much a healing process as a process exactly opposite to that described above. (1)

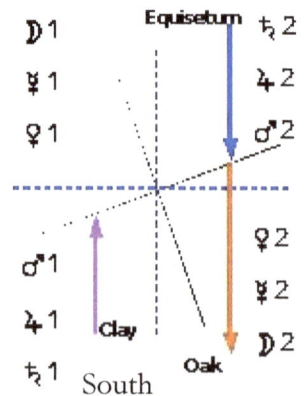

148

With regards the kidneys, Equisetum works by stimulating the activity of the Internal Astral body and Cosmic Substance against too strong a World Astral invasion. This then allows the internal astral to regain its control over the Internal Etheric and Earthly Forces, which brings the overly strong lunar processes into order.

We also need to remember the important role clay plays in the strengthening of the Cosmic Forces, and ensure that there is an adequate supply of this, to push through the strong lunar forces, especially in sandy soils.

We can go one step further. Healing is not based on the microscopic changes in tissues and cells, but on a knowledge of the larger connections; this must also be our attitude to the plant nature. And since plant nature is in this respect simpler than that of the animal or man, so its healing is a more general process and when sick it can be healed with a kind of "cure-all" remedy. If this were not so, we should often be in a fix with regard to plants, as we are with animals, though not with human beings. For a man can tell us where he feels pain. Animals and plants cannot; and it is fortunate that. here the curative process is almost the same for all plants. A large number of plant diseases (although not all of them) can really be arrested as soon as they are noticed by a rational management of our manuring - namely in the following way: We must then add calcium to the soil by means of the manure. But it will be of no use if the calcium is not applied in a living condition. If it is to have a healing effect it must remain within the realm of the living. Ordinary lime or the like is of no use here. Now we have a plant which is very rich in calcium - seventy-seven per cent, of its substances is calcium albeit in very fine distribution. This is the oak and more especially its bark. In the bark we have something which is at an inter-mediate stage between plant and living earth. You will remember what I said to you about the kinship between bark and live earth. For calcium as required in this connection the calcium structure in the bark of the oak is almost ideal. Calcium in a living state (not dead, though even then it has an effect) has the property which I have already described to you: it restores order where the etheric body is working too strongly so that the astral element is prevented from reaching the organic substances. Calcium kills (damps down) the forces of the etheric body and so sets free those of the astral body. This is characteristic of all limestone. But if it is necessary for an over-powerful etheric element to be damped down and contracted in a regular way - not suddenly nor jerkily so that shocks are produced - but in a steady and orderly fashion, we should use calcium in the particular form in which it is to be found in the bark of the oak tree.

Powdery Mildew

Fungal attacks, such as Powder Mildew, can occur from the opposite direction. Human phenomena such as diphtheria are especially able to teach us about certain subtleties in the plant organism. Such diseases should be studied more precisely, if only for the sake of discovering remedies.

In another context I have indicated that the child's acquisition of speech is accompanied by various organic processes. While he is learning to speak, and therefore while something special is taking place in his breathing organism, something also occurs polarically in his circulatory organism, which also receives into itself the metabolic processes. I also pointed out, how what at puberty appears in a reciprocal relationship of the human being to the outer world, takes place inwardly in learning to speak. Thus this push of the astral body, which at puberty takes place from within the human being outward, takes place from below upward in the capacity for acquiring speech. *(from the metabolic towards the nerve sense)* So here we have an astralizing process, and we will be able to see clearly that an interaction occurs where the respiratory and circulatory systems meet (see drawing). The astralizing process working from below upward (yellow) encounters the developing organs of speech working from above downwards (red). In this

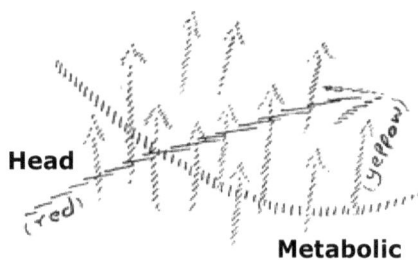

encounter the organs of speech become stronger in their capacity for speech. It is what is taking place simultaneously below *in the metabolism,* that especially interests us here: this tends to work upward. The whole process is one from below upward (yellow arrows). Now, if the astrality presses upward too strongly while the child is learning to speak, we have a predisposition to diphtheric conditions. It is certainly important to pay proper attention to this.

Let us now consider the outer earthly *forces* process, *we see with plants,* that has a certain selective affinity for the process I have just described. Let this be the surface of the earth . In a plant that behaves appropriately in relation to the cosmos, the earth plays a part in the *nerve sense* formation of its roots. With growth the influence of the earth diminishes and the *Cosmic Substance* influence becomes stronger and stronger, unfolding especially in the blossoms (see drawing, red). What develops here is a kind of external astralizing of the blossom, which then leads to the formation of fruit. If this process, which ought to occur in the normal course of the world processes, takes place below , it can only insert itself into the water, and we have what I have just called "dysentery of the earth."

But we can also have another situation: What takes place when a plant develops properly — the blossom unfolding always a little above the earth's surface — *and can bring the Cosmic Substance* right on the earth's surface (see drawing below, red). Then fungi arise; this is the basis for fungus formation, *such as Powdery Mildew.*

And now you will begin to guess that, if fungi arise from such a special astralizing process, the same process must take place from below upward when, as in human diphtheria, this remarkable astralization occurs in the human head. This is actually the case. Hence you find in diphtheria the tendency to fungoid formations. It is most important to consider this tendency to fungoid formations in diphtheria, and it will also show you that a truly occult process is taking place there. Everything external is really only a sign that irregular astral currents, of *Cosmic Substance from the metabolism towards the head,* are prevailing within the human being.

But when, as here, the processes work so deeply into the organism, much more will naturally be achieved by trying to find the specific remedy with which to oppose the particular process at work. One should try intermediate potencies of **cinnabar** - *Mercury Sulphate*. In cinnabar we will find effects that counteract all the phenomena I have mentioned. Cinnabar expresses this even in its outer appearance. If we acquire a sound understanding of such things we will recognize that cinnabar through its vermilion color is something that in a certain way brings to expression this activity opposed to the fungoid process. That which is approaching the colorless can become fungoid. While too strong an astralization of the earth's surface plays a part in the formation of fungi, in cinnabar there is a counter-reaction to this astralization and thus this reddening. *Mercury a brother of Zinc is an element that strengthens the Etheric within the Physical body and so pushes off a too strong astral working*. Wherever a reddening appears in natural processes, we find a powerful counter-effect to the astralization process. You could express this in a moral formula: "The rose in blushing works against astralization." These domains of pathological-therapeutic study are really interconnected in a certain way. They guide us into this peculiar relationship of the ego and astral body to the other organs, to their laying hold of organs, to their emancipation from organs, or to manifestations of the excessive working of the astral from below upward *in the human, which for the plant is from above downwards*. (2)

This process is very similar to that occurring in some pest attacks.

Now I am going to tread on very thin ice and take an example very near home. I am going to talk about the nematode of the beetroot. The outer signs of this disease are a swelling of root fibers and limpness of the leaves in the morning. Now we must clearly realise the following facts: The leaves, the middle part of the plant which undergo these changes, absorb cosmic *substance* influences that come from the surrounding air, whereas the roots absorb the *cosmic* forces which have entered into the earth and are reflected upwards into the plant. What, then, takes place when the nematode occurs? It is this: The process of absorption of *Cosmic Substance* which should actually reside in the region of the leaves has been pressed downwards and embraces the roots.

Thus if this (Diagram No. 10) represents the earth level, and this the plant, then in the plant infested with the nematode the *astralised Cosmic Substance* forces which should be active above the horizontal line are actually at work below it. What happens is that certain cosmic *substance* forces slide down to a deeper level; hence the change in the external appearance of the plant. But this also makes it possible for the parasite to obtain under the soil (which is its proper habitat) those *astralised* cosmic *substance* forces which it must have to sustain It (the nematode is a wire-like worm). Otherwise it would be forced to seek for these forces in the region of the leaves; this, however, it cannot do as the soil is its proper environment. Some, indeed all, living beings can only live within certain limits of existence. Just try to live in an atmosphere 70 degrees above or 70 degrees below zero and you will see what will happen. You are constituted to live in a certain temperature, neither above nor below it. The nematode is in the same position. It cannot live without earth and without the presence of certain cosmic *substance* forces brought down into it. Without these two conditions it would die out.

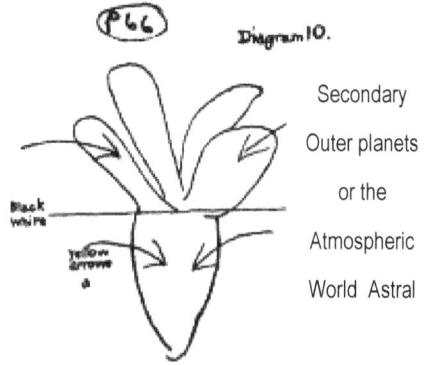

Every living being is subject to quite definite conditions. And for the particular beings with which we are dealing, it is important that cosmic forces should enter the earth, forces which would ordinarily display themselves only in the atmosphere around the earth. *If we change the energetic environment the pests and disease will go elsewhere.*

Animals

Now why do I say that the "agricultural-individuality" stands on its head?

153

I do so because the air, vapours and warmth which are in the immediate neighbourhood of the soil and from which both man and the plants derive air, moisture and warmth - all this corresponds to the abdominal organs in the human body. On the other hand every thing that takes place within earth, under the soil, affects the general growth of plants in the same way as our head affects our organism - especially in childhood, but also throughout the whole of our life. Thus there is a constant and very living interplay of supra-terrestrial and sub-terrestrial activities.

To what I said about the "belly" being above the Earth and the "head" being under the Earth, belongs an understanding of *the internal workings within* the animal organism.

In the animal, the threefold organism is not so sharply defined as it is in man. The animal has a system of nerve and senses and a metabolic and limb system. These are clearly divided, the one from the other. But in many animals the limits of intermediate rhythmic system are indefinite; both nerves and senses system and metabolic system trespass upon the limits of the rhythmic system. We should therefore choose other terms when we speak of animals. In man one is quite right in speaking of a three-fold organism: but in the case of animals one ought to speak of the nerve and senses system as being

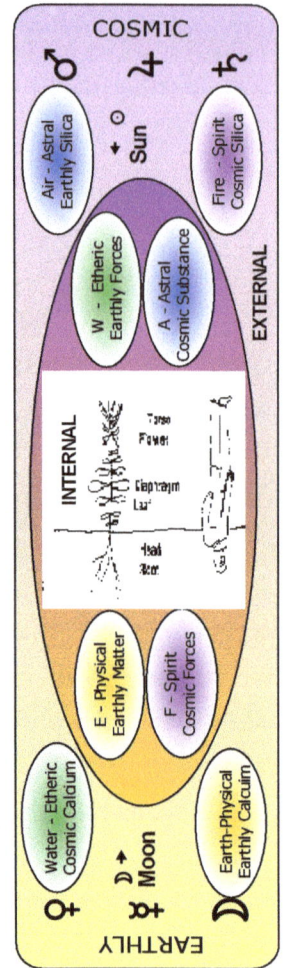

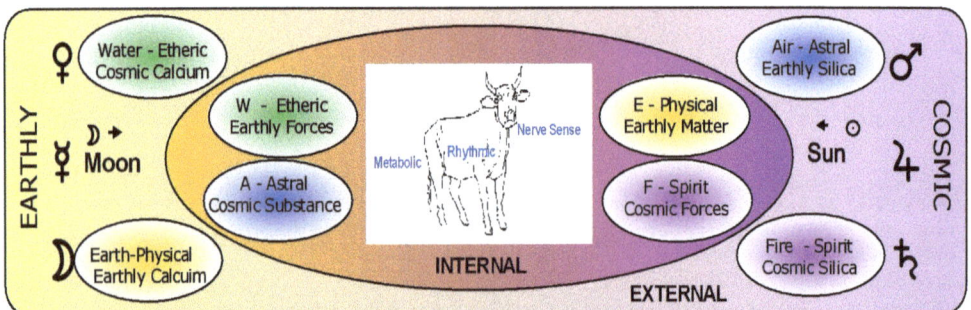

localised primarily in the head, *and working backwards throughout the whole body,* and of the metabolic and limb system as being in the hind quarters and limbs but at the same time *working forwards* throughout the whole body. In the middle of the body the metabolism becomes more rhythmical as does also the nervous system, and there both flow into one another. The rhythmic system has a less independent existence in the animal. Rather the opposite poles become indistinct as they merge into one another. We should therefore speak of the animal organism as being twofold, the extremes interpenetrating at the middle. In this way the animal organization arises.

Now *internally,* all the substances contained in the 'head' system - I am speaking of animals, but the same is true of man *and plants* - are of Earthly Substance. Even in the embryo, Earthly Substance is led into the head system. The embryo must be so organised that its head receives its matter from the earth. In the head, therefore we have Earthly Matter. But the substances which we bear in the metabolic and limb organisation, those which permeate our intestines, our limbs, our muscles and bones, etc., these substances do not come from the earth, but from what has been absorbed from the air and warmth above the Earth.- It is Cosmic *Substance.* This is important. When you see an animal's claw, you must not think of it as having been formed by the food which the animal has eaten and which has gone to the claw and been deposited there. This is not the case. It is cosmic matter taken up through the senses and the breathing. *The etheric aspect of* what the animal eats serves only to stimulate its powers of movement , so that the cosmic substance can be driven into the metabolic and limb organisation, and can be driven into the claw and similarly distributed throughout the whole organism.

With Forces (*as opposed to Substances* it is the other way round. Because the senses are centred in the head and take in impressions from the cosmos, the Forces in the head are Cosmic in nature. To understand what happens in the metabolic and limb organisation, you need only think of walking, which means that the limbs are permeated with earthly gravity: the Forces are Earthly ones. Thus the limb system contains Cosmic Substances permeated by Earthly Forces. It is extremely important that the cow or the ox, if used for working, should be fed so as to absorb the greatest possible amount of Cosmic Substance, and that the food which enters its stomach should produce the necessary *Earthly Force* strength to lead this Cosmic Substance into its limbs, muscles and bones. *Thus the Etheric inspired Earthly Forces provides an enlivening levity to the Cosmic Substances' drive to manifestation .*

It is equally important to realise that the Earthly Substances in the head have to be drawn from the food, which has been worked upon in the stomach, and is led into the head. In this sense, the head relies upon the stomach in a way in which the big toe does not, and we must realise quite clearly that the head can only work upon this nourishment which comes to it from the metabolism, if it can at the same time draw in sufficient Cosmic Forces. If, therefore, animals instead of being left in stuffy stables where no cosmic forces can reach them, are led into meadows and given every opportunity of entering into relation with their environment through the perceptions of their senses, then we may see results such as appear in the following examples.

Imagine an animal standing in a dark and stuffy stable before its manger, the, contents of which have been measured out by human "wisdom". Unless its diet is varied, as it only can be out-of-doors, this   animal will show a very great contrast to one which seeks out its food with its sense of smell, guided by this organ in its search for Cosmic Forces, seeking and finding its nourishment by itself and developing its whole activity in doing so. An animal that is fed from a manger will not show immediately how devoid it is of Cosmic Forces, for it has inherited a certain amount of them. But it will breed descendants to whom these Cosmic Forces are no longer transmitted. Such an animal will become weak, beginning from the head, i.e. it will not be able to nourish its body because it cannot take in, the necessary cosmic substances which should come in. This will show you that it is not enough simply to say: "This kind of fodder for one case, that for another". Rather one must have a clear idea of the value for the animal's whole organisation that such and such methods of feeding have.

Cowhorns

Now following the trend, *we saw in compost,* we can take a further step. Have you ever wondered why it is that cows have horns, while certain other animals have antlers? It is a very important question. Yet what science has to say about it is quite one-sided and based on externals. Let us consider why cows have horns. I said that the forces within a living organism need not always be directed outwards, but can also be directed inwards. Now imagine an organic entity possessing these two sets of forces, but which is unformed and lumpish in build. The result would be an irregular, ungainly being. We should have curious looking cows if this were the case. They would all be lumpish and unformed, with rudimentary limbs as at an early embryonic stage. But this is not how a cow is constructed. A cow has horns and hoofs. Now what happens at the points where horns and hoofs grow? At these points an area is formed from which the organic formative forces, *moving outwards from the metabolism,* are reflected inwards in a particularly powerful way. There is no communication with the outside as in the case of the skin or hair; the horny substance *of the horn* blocks the way for these forces to the outside. This is why the growth of horns and claws has such a bearing upon the whole form of the animal.

Things are quite different in the case of antlers. Here the streams of forces, *coming from the metabolism* are not led back into the organism, but certain of them are guided for a short distance out of the organism; there must be valves, as it were, through which the streams localised in the antlers (we can speak of streams of 'force', just as we can speak of streams of air or liquid) can be discharged. A stag is beautiful because it stands in intense communication with its environment by reason of its sending outwards streams of *metabolic* forces; by this it lives within its environment and takes up from it everything which works organically in its nerves and senses. Hence the nervous nature of the stag. In a certain respect all animals which have antlers are suffused with a gentle nervousness. This is clearly to be seen in their eyes.

The cow has horns, in order to reflect inwards the astral *inspired cosmic substance* and etheric inspired earthly forces, *coming first from the metabolism*, which then penetrate right *back* into the metabolic system, so that increased activity in the digestive organism arises by reason of this radiation from horns and hoofs. If one wants to understand foot-and-Mouth disease, i.e. the retroaction from the periphery to the digestive tract, one must know of this connection. Our remedy for Foot-and-Mouth disease is based on the recognition of this. In the horn, therefore, we have something which by its inherent nature is fitted to reflect the living etheric and astral streams into the inner life organs. The horn is something which radiates etheric life and even the astral element. Indeed, if you were able to enter into the cows belly, you would smell the current of etheric-astral life which streams *back* from the horns: and the same thing is true of the hoofs.

Now this gives us a hint as to the measures we may recommend for increasing the effectiveness of ordinary stable manure. What is ordinary stable manure really? It is foodstuff which the animal has taken in and which up to a certain point has been assimilated by its organism, thereby stirring into activity certain dynamic forces in the organism. Its main use has not been to increase the amount of substance in the organism, for after having had its effect, it is excreted. It has become permeated with astral *cosmic substance* and etheric *earthly force* elements. The astral element has filled it with nitrogen-bearing forces and the etheric element with oxygen-bearing forces. The substance which emerges as dung is permeated with these forces. Imagine now: We take this substance and pass it into the soil in some form or other (the details will be dealt with later). Thus we add to the soil an etheric-astral element whose proper place is in the belly of the animal, where it produces forces of a plant-like nature. For the forces which we produce in our digestive tract are of a plant-like nature. We should be extremely thankful that we get such a residue as dung, for it carries etheric *earthly forces* and astral *cosmic substance* forces from the interior of the organism out into the open. These forces remain with it, and it is for us to keep them there. In this way the dung will act in a life-giving and also astralising way on the soil, not only on the water element in it, but especially on the solid element. It has the power to overcome what is inorganic in the earthly element. Now what is passed over to the soil will necessarily, of course, lose the form it originally had when taken in as food, for it has to go through an inner organic process in the metabolic system. There it enters upon a phase of decomposition and dissolution. But it is at its best just at the point where it begins to dissolve through the workings of its own astral and etheric elements. It is then that the parasites, the micro-organisms make their appearance. They find a good feeding-ground in which to develop. This is why the theory arose that these parasites are themselves responsible for the virtues in the manure. But they are only indications of the condition of the manure. If we think that by inoculating the manure with these bacteria we shall radically improve its quality, we are making a complete mistake. Externally there may seem at first to be an improvement, but in reality there is none.

Animal Husbandry

Now, to go further: in any given region of the earth there is not only a particular vegetation but also certain animals live there. For reasons which will appear later on, we need not consider human beings for the moment. It is one peculiar fact, and I should be glad to see this put to experimental test as I am quite sure that such a test would confirm it. This fact is that the right quantities of cows, horses and other live-stock on a farm will supply just the necessary amount of manure for the farm to restore to it what has been discharged into "chaos". Moreover the right proportion of horses, cows and pigs will yield the right proportions in the mixture of manures. This is because the animals eat the right proportion of the plant substances yielded by the soil, and because in the course of their organic processes they produce as much manure as is needed to be given back to the soil. And, though it cannot be strictly carried out, I would say that manure of any kind introduced from outside can only be regarded as a curative substance for a farm that has become diseased.

A farm is only healthy if it can supply itself from the manure yielded by its own animals. This of course entails the development of a real knowledge of how many animals of a given sort are necessary for a given farm. But this will be found out as soon as some knowledge returns to us of the inner forces in Nature. To what I said about the "belly" being above the Earth and the "head" being under the Earth, belongs an understanding of the animal organism.

Thus if we imagine ourselves to have picked up the animal, turned it round and set it upside down with its head in the earth we shall have the position invisibly taken by the "agricultural-individuality". The consideration of this formation of the animal enables us to see a relation between the manure produced by the animal and the needs of the earth in which the plants grow which serve as food for the animal. For you will remember that the cosmic forces which act in a plant are guided upwards through it from inside the earth. If, therefore, a plant is particularly rich in these cosmic forces, and an animal eats it, then the manure which this animal excretes will be particularly well-suited to the soil on which the plant grows. Thus if we learn to grasp the forms of things we shall see in what sense an agricultural unit, or farm is a "self-contained individuality" (or as we have called it an agricultural-individuality") only we have to include in it the necessary livestock.

But we must go a step further. What is actually - contained, in the Head? Earthly Substance. If you take out the brain, the noblest part of an animal, you will have before you a piece of Earthly Substance. The human brain also contains Earthly Substance. But in both the Forces are Cosmic. What is the human brain for? It observes as a support for the Ego. The animal, let it be remembered, has as yet no Ego; its brain is only on the way to Ego-formation. In man it goes on and on to the complete forming of the Ego. How then did the animal's brain come into existence? Let us look at the whole organic process. All that which eventually manifests in the brain as earthly substance has simply been "excreted, *and deposited*, from the organic process. Earthly Substance has been excreted in order to serve as a base for the Ego *and Cosmic Forces*. Now the process of the working-up of the food in the digestive tract and metabolic and limb system produces a certain quantity of Earthly Substance which is able to enter into the head and to be finally deposited as Earthly Substance in the brain. But a portion of the food stuff is eliminated in the intestine before it reaches the brain. This part cannot be further transformed and is deposited in the intestine for ultimate excretion.

We come here upon a parallel which will strike you as being very paradoxical but which must not be over-looked if we wish to understand the animal and human organisations. What is brain matter? It is simply the contents of the intestines brought to the last stage or completion. Incomplete (premature) brain-excretion passes out through the intestines. The contents of the intestines, are in their processes, closely akin to the contents of the brain. One could put it somewhat grotesquely by saying that that which spreads itself out in the brain is a highly advanced dung-heap. And yet the statement is essentially correct. By a peculiar organic process, dung is transformed into the noble matter of the brain, there to become the foundation for the development of the ego. In man the greatest possible quantity of intestinal dung is transformed into cerebral excrement because man bears his ego on the earth. In animals the quantity is less. Hence there remain more forces in the intestinal excrement, of an animal which we can use for manuring. In animal manure, there is therefore more of the potential ego element, since the animal itself does not reach ego - hood. For this reason animal dung and human dung are completely different.

Animal dung still contains ego-potentiality. In manuring a plant, we bring this ego-potentiality into contact with the plant's root. Let us draw the plant in its entirety (Diagram 16). Down here you have the root; up there the unfolding leaves and blossoms. And as above, in the leaves and blossoms, the Astral element (red), *Cosmic Substances'* *'relative'*, is acquired from contact with the air, so the ego-potentiality (orange) *Cosmic Forces' 'relative'*, develops below in

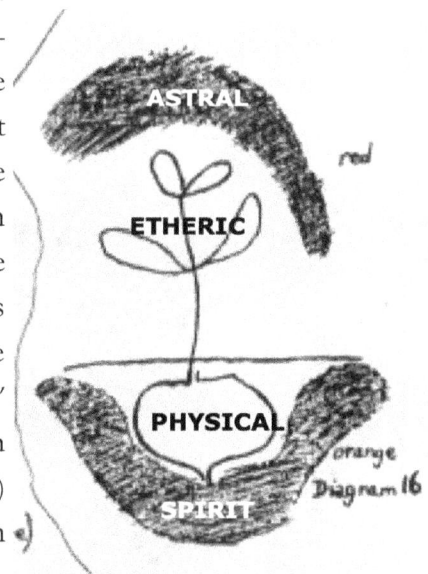

Diagram 16

161

the root through contact with the manure.

The farm is truly an organism. The astral element is developed above, and the presence of orchard and forest assists in collecting it. If animals feed in the right way on the things that grow above the earth, then they will develop the right ego-potentiality in the manure. If they produce, this ego-potentiality, it will work on the plant from the root, will cause it to grow upwards from the root in the right way according to the forces or gravity. It is a wonderful interplay, but in order to understand it one must proceed step by step.

To begin with the root. The root generally develops in the soil and through the manure it becomes permeated with ego-potentiality which it absorbs. This absorption is determined and aided if the root can find in the right quantities salts in the soil around it. Let us assume that we are considering the nature of these roots merely from the point of view of the foregoing reflections. Then we shall suggest that roots are the food which, when it is absorbed into the human organism, will find its way most easily to the head by way of the digestive process. We shall therefore provide a diet of roots where we require to give the head material substances to

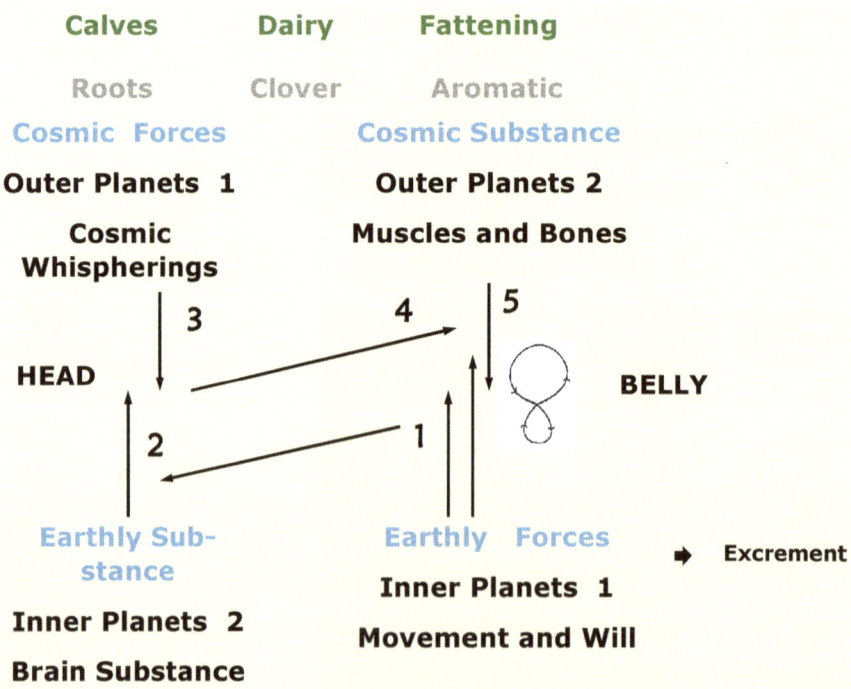

Calves	Dairy	Fattening
Roots	Clover	Aromatic
Cosmic Forces		Cosmic Substance

Outer Planets 1   Outer Planets 2

Cosmic   Muscles and Bones
Whispherings

| 3   4 |5

HEAD   BELLY

2

Earthly Sub-   Earthly Forces   ➤ Excrement
stance
                          Inner Planets 1
Inner Planets 2
                          Movement and Will
Brain Substance

enable the cosmic forces which work through the head to exercise their plastic activity. Now imagine someone saying to himself: "I must give roots to this animal which requires Earthly Substance in its head in order to stimulate its sense-connections with the cosmic environment". Does not this immediately suggest the calf and the carrot. A calf eating carrots portrays this whole process. The moment something like this is put forward and you know how things really are and their true connections, you will know immediately what is to be done. It is simply a matter of realising how this mutual process arises.

But let us proceed to the next stage. Once the calf has eaten the carrot, once the *Earthly* Substance really has been introduced into the head, the converse process must be able to begin, i.e. the head, on its part, must begin to work with forces of volition thus begetting within the organism *Cosmic Substance* forces which can be worked into it. It is not enough for the "carrot dung" to be deposited in the head; from what is deposited and in the course of disintegration, streams of force must come and enter the rest of the organism. In short, there must be a second food substance which will enable one part of the body which has already been fed (in this case the head) to work in the right way on the rest of the organism.

Well, I have given the animal the carrot fodder. And now I want the animal's body to be permeated with the forces *of Cosmic Substance,* which are *first* developed from the head. For this, as a second fodder, we need a plant with a spindly structure, the seed of which will have gathered into itself these "spindly" forces. We immediately think of flaxseed (linseed) or something similar. If you feed young cattle on carrots and linseed - or carrots and fresh hay (which is equally suitable) - this will bring into full operation the forces already, latent in the animals. We should therefore try to give young cattle food which promotes, on the one hand, the *Cosmic* Forces of ego-potentiality, and, on the other, the complementary streams of astral force working from above downwards *as Cosmic Substance* : For the latter purpose, those plants are especially suitable which have long, spindly stems and as such have been turned into hay. Just as we have looked into this concrete case, so we must approach

163

Agriculture as a whole: of every single thing, we must know what happens to it when it passes either from the animal into the soil, or from the plant into the animal.

Now coming from the other direction, it is quite true that what one eats is important, but the greater part of it is not there for the purpose of being taken into the body and deposited there as substance. This greater part has to give over to the body the *Earthly* Forces which it contains in itself and thus stimulate the body into activity. The greater part of what is taken up as *food* substance in this way is eliminated again from the body. What matters, therefore is not whether a certain weight of matter in certain proportions undergoes digestion, but whether we are able to take up in the right way, with the food we eat, the active forces therein. For we need these active *Earthly* Forces when we walk or work, or even more when we use our arms. On the other hand, that which the body needs in order to fill up, to enrich itself, as it were, with substance (the substance being continually discarded and renewed during the course of every seven or eight years) is absorbed for the most part through the sense organs, the skin and the breathing, in a highly attenuated state, as *Cosmic Substance,* and only *then* becomes densified in the organism. The body absorbs it from the atmosphere, densifies and hardens it, so that for instance it can be cut off as hair and nails. The schematic formulation: "Food taken in, passage through the body, wearing away of the nails, peeling of the skin, etc" is quite wrong. It should run: "Breathing, highly rarefied absorptionof *Cosmic Substance,* through the sense organs (even through the eyes), passage through the organism, excretion." What is absorbed through the digestion on the other hand *as Earthly Forces* becomes important because its "inner ' mobility" (Regsamkeit) is set free, just as when fuel is burned. It introduces into the body those forces which open the way for the will to act in the body. If these things are to be rightly handled, it is necessary to gain insight into the mode of activity of substances *(physical)* , and forces *(etheric),* the dynamic *(astral)* and of the spiritual too in every part of agriculture. A child who does not know what a comb is for, will bite into it or otherwise misuse it. In the same way we shall make quite a wrong use of things if we do not understand their essential being and their specific functions.

Dairy Cattle

Let us pursue the subject yet further. Let us take the case of an animal which should become particularly strong in the middle region (where the head or nervous organisation tends to develop in the direction of breathing and the metabolic organisation tends to have a rhythmic character). Which animals have to be strong in this particular region? They are the milch animals. The secretion of milk shows that the animal in question is strong in this region. The point to observe here is that the right co-operation should take place between the *Sun* current going from the head backwards (mainly a streaming of forces) and the *Moon* current going from the animal's hind-quarters forward (mainly a streaming of substance.) If these two currents co-operate and intermingle in the right way, the result will be an abundant supply of rich milk. For good milk contains substances prepared in the metabolic system and which, without having entered into the sexual system, have become akin to it. It is a sexual process within the metabolic system. Milk is simply a sexual secretion on another level. It is a substance, which, on its way to becoming sexual secretion, is penetrated and transformed by the forces working from the head. The whole process can be seen quite clearly.

Now for processes which should arise in this way, we must choose a diet which will work less powerfully towards the head than do roots which contain ego-potentiality; neither may the diet, since it is to be connected with the sexual system, contain too much of the astral element, i.e. of that which goes towards the blossom and fruit of the plant. In short, if we wish to find a diet that will produce milk, we must choose the part of the plant which lies between blossom and root, i.e. the green and leafy part. If we wish to bring about an increase in the milk supply of an animal whose milk production we have reason to believe could be increased we shall certainly reach the desired end if we proceed as follows: Suppose I have a cow and feed it with green fodder. I take plants in which the process of fruit-formation has been developed within the process of leaf-formation. Such, for example, are the pod-bearing or leguminous plants and especially the clovers. In clover, the would-be fruit develops as leaf and foliage. A cow that is fed in this way will perhaps not show much result of it; but when the cow comes to calve, the calf will grow into a cow that yields good milk. The effects of reformed foddering usually need a generation in which to show themselves.

Suppose now that we wish to consider the flowering and fruiting part of the plant. And we must go further, and observe what is fruit-like in the rest of the plant. This recalls a feature of plant-life that always delighted Goethe, namely the fact that the plant has throughout its whole body the tendency towards what is normally specialised at certain parts. With most plants we take the seed which has formed from the blossom and place it in the earth in order to produce more plants. But we do not do this in the case of the potato. Here we use the eyes of the tubers. This is the fruiting part of the potato plant, but, like many processes in Nature, it is not carried out to the end. We can, however, heighten its activity by a procedure which bears an external resemblance to combustion. For instance, if you "cossette" (chop up into thin straws) roots or tubers and dry the "cossettes" for fodder, the stuff will be enormously strengthened in its activity and brought a stage nearer to the fruit stage if you spread it out in the sun and allow it to steam a little.

QUESTION: What does Spiritual Science say on the subject of preserving foodstuffs by acidification in general?

ANSWER: If we use salt-like materials at all in this process it does not make much difference whether the salt is added at the moment of eating, or whether it is used in the preparation of the fodder. In the case of fodder that contains too little salt to carry the food stuffs to those parts of the organism where they should work, souring is the right procedure to adopt. Take the case of turnips. These, as we saw, are particularly fitted to work upon the head-organisation. They are, therefore, an excellent food for certain animals, especially for young cattle. If, however, it be noticed that the young animals shed their hair too soon and too much, their fodder should be salted because this means that the food is not being deposited in sufficient quantities in those parts of the organism where it is needed. Salt is tremendously effective in carrying food to the part of the organism where it is needed and will work.

QUESTION: What view does Spiritual Science take on the subject of souring of the leaves of sugar-beet and other green plants?

ANSWER: The great thing here is to find a certain optimum and not go beyond it by adding too much salt, because salt is the part of food which more than any other remains what it is once it is inside the organism. The organism in general, in the case of animals and even more so of human beings, is so constructed as to submit everything it absorbs to the most varied changes. It is an error to think that the albumen which goes into our stomachs remains the same as it was before we ate it. It must first be changed into a completely lifeless substance and then changed back again by means of the etheric body into specifically human (or in the case of animals specifically animal) albumen.

Everything that enters into an organism must be changed. This applies even to warmth. Suppose that this (left) is a living organism and this the warmth in the environment. Now assume you have a piece of dead wood (right) which, it is true, comes from a living organism but is already dead. It is likewise surrounded by warmth. Now when the warmth enters into the living organism, it does not simply go a little way in and remain what it is; the organism immediately transforms it into a warmth of its own, and it could not do otherwise. Whereas when the warmth penetrates into the dead wood it remains exactly the same kind of warmth as exists outside in the mineral earth. The moment warmth penetrates into us unchanged as it does into the piece of wood, we catch cold. Nothing that comes into the living organism from outside may remain what it is* it must immediately be changed into something else. This process takes place to the least extent in salt. No great harm, therefore, will be done by using salt for the

preserving of food-stuffs so long as you do it carefully and do not put in too much. The mere sense of taste will reject it. If it is necessary for the preservation of food-stuffs this shows that up to a point it is the right process to adopt.

QUESTION: Do you recommend souring fodder without salt?

ANSWER; That is too advanced a process. It is a super-organic process (self-fermentation) and can in certain circumstances be extremely harmful.

Cooking

Practices like this are based upon a deep and wonderful instinct. We can ask: how did men first come to cook their food? Men began to cook their food because they gradually discovered that what develops during fruit formation is mainly due to processes akin to cooking, viz. burning, warming, drying and evaporating. All these processes tend to make the fruit and seed and in-directly "the other parts of the plant, especially the higher parts, more fitted to develop the forces that are necessary to the metabolic and limb system in the animal. Even uncooked the blossom and fruit of a plant work on the animal's metabolic and digestive system and primarily through the forces they develop, not through their substance. For it is the forces of the earth which are needed by the metabolic and limb system, and in the measure in which it needs them, it must receive them. Take the case of the animals which pasture on steep mountain sides. Unlike those in the plains, they climb about under difficult conditions owing to the fact that the ground is not level. There is all the difference for those animals between level and slanting ground. They require food that will develop those forces in limb and muscle which are energised by the will. Otherwise they would not be good for either labour, milking or fattening. It is therefore important that they should eat plenty of those aromatic mountain plants in which blossom and fruit have undergone an additional treatment by the sun, resembling a process of natural cooking.

But similar results can be achieved and strength given to muscle and limb by artificial methods - roasting and boiling, etc. Flower and fruit are most suitable for this, especially of those plants which, from the beginning develop towards fruiting and do not waste their time, as it were, in growing foliage.

Fattening Animals

This brings us to the question of the fattening of animals. Here we must say we should regard the animal as a kind of sack to be filled as full as possible with Cosmic Substance, *supported by Earthly Forces*. A fat pig is really a most heavenly animal! It's fat body, apart from its system of nerves-and-senses, is made up entirely of cosmic, not of earthly substance. The pig needs the food which it enjoys so much in order to fill itself with cosmic substance, which it absorbs on all sides and then distributes throughout its body. It must take in this substance which has to be drawn from the cosmos, and distribute it. And the same is true of all fattened animals. You will find that animals will fatten best on the part of the plant which tends towards fruit-formation, and has been heightened in its activity by cooking or steaming. Or, if you give them something which has in it an enhanced fruit-process, for instance turnip, which belongs to a species in which this process has been enhanced and which has become larger through long cultivation. In general, the best kind of food for fattening cattle is that which will at least help to distribute the cosmic substance, i.e. the part of the plant which tends to fruit-formation - and which has in addition received the proper treatment. These conditions are in the main fulfilled by certain kinds of oil cakes and the like. But we must also see to it that the animal's head is not entirely neglected and that in this fattening treatment a certain amount of earthly substance is introduced there. The fodder just mentioned needs to be complemented by something for the head, though a smaller quantity, as the head does not require so much. In fattening an animal, we should therefore add a small quantity of roots.

Now there is a substance which as substance has no particular function in the organism. In general, one can say that roots have a function in connection with the head, blossoms in connection with the metabolic and limb system, and leaf and stem in connection with the rhythmic system within the human organism. There is, now, a substance that can aid the whole animal organism, because it is related to all its members. This substance is salt. And as of all the ingredients in the food of both human and animals, salt is the least in quantity, we can see it is not how much we take which matters, but what we take. Even small quantities of substance will fulfill their purpose if they are of the right kind.

The essential thing about nourishment is that forces should be developed in the body. Whether the animal develops enough forces to enable it to take in and transform the substances in the atmosphere depends upon whether it absorbs its food in the right way. To make a comparison. If you want to put on a close-fitting glove you don't do it by squeezing your fingers into it. You first enlarge the glove with a stretcher. In the same way we must bring elasticity into those forces which are to take out of the atmosphere what is not produced by food. Through the food the organism is stretched and thereby enabled to take in more of what it needs from the atmosphere. This may even lead to hypertrophy if too much food is taken in. This has to be paid for by a shortened life span. The middle course must be found between the maximum and minimum. (1)

Further Considerations

What here plays in — the fact that man feels himself as a being constituted in the way I have described — this at first remains for the ordinary consciousness of today below in the unconscious. There, certainly, it is already present; and there it emerges as a kind of mood, a kind of life-mood of man. But it is spiritual vision alone that brings it to full consciousness, and I can only describe this spiritual vision to you thus: The man who knows from present-day initiation-science this secret of the human being, namely that the head is the most important, the most essential organ which needs *Earthly* substance with *Cosmic* forces; who knows further that the most essential thing in the system of limbs and metabolism is *Cosmic* substance which needs *Earthly* forces — the forces of gravity, of balance, and the other *Earthly* forces in order to exist; who can thus penetrate with spiritual vision into this secret of the human being and who then turns his gaze back to this human, earthly existence — this man must acknowledge himself as a tremendous debtor to the world. For he must admit that in order to maintain his human existence he requires certain conditions; but through these very conditions he becomes a debtor to the earth. He is continually withdrawing something from the earth. And he finds himself obliged to say that the *Cosmic* substance, which as man he bears within himself during earthly existence, is actually needed by the earth. **When man passes through death, he should in fact leave this *Cosmic* substance behind him for the earth,** for the earth continually needs *Cosmic* substance for its renewal. But this man cannot do, for he would then be unable to traverse his human path through the period after death. He must take this *Cosmic* substance with him for the life between death and a new birth; he needs it, for he would disappear, so to speak, after death, if he did not take this *Cosmic* substance with him. *Cosmic Substance carries the dross of our Astral body as we have moved through life.*

Only by carrying this *Cosmic* substance of his limb-metabolic system through the gate of death can man undergo those transformations which he must there undergo. He would be unable to meet his future incarnations if he were to give back to the earth this *Cosmic* substance which he actually owes to it. He cannot do this. He remains a debtor. And this is something which there is no means of bettering as long as the earth remains in its middle period. At the end of earth-existence things will be otherwise.

It is indeed the case, my dear friends, that one who beholds life with Cosmic vision has not only those sufferings and sorrows — perhaps also that happiness and joy — which are offered by ordinary life, but, with the beholding of the *Cosmic,* cosmic feelings, cosmic sufferings and joys, make their appearance. And initiation is inseparable from the appearance of such cosmic suffering as, for example, the fact that one has to admit: Simply because I must maintain my humanity I must make of myself a debtor to the earth. I cannot give to the earth what I really should give if, in a cosmic sense, I were to act with complete rectitude.

Matters are similar as regards the *Earthly* substance which is present in the head. Because throughout the entire course of earth-life *Cosmic* forces are working in the Earthly substance of the head, this head-substance becomes estranged from the earth. Man must take away from the earth the substance for his head. But he must also, in order to be man, continually imbue this substance of his head with extra-terrestrial *Cosmic* forces. And when the human being dies, this is something extremely disturbing to the earth, because it must now take back the substance of the human head which has become so foreign to it. When the human being passes through the gate of death and yields up his head-substance to the earth, then this head-substance — which is entirely *Cosmicized,* which bears within itself what results from the *Cosmic Forces* — does in fact act as a poison, as a really disturbing element, in the totality of the life of the earth. When man sees into the truth of these matters, he is obliged to say to himself that the honest thing would be to take this substance with him through the gate of death, for it would in fact be much better suited to the spiritual region which man traverses between death and a new birth. He cannot do this. For if man were to take this *Cosmicized* earth-substance with him, he would continually create something adverse to all his development between death and a new birth. It would be the most terrible thing that could happen to man if he were to take this *Cosmicized* head-substance with him. It would work incessantly upon the negation of his spiritual development between death and rebirth.

One must therefore acknowledge, when one sees into the truth of these things, that here, too, man becomes a debtor to the earth; for something for which he is indebted to the earth but has made useless for it, this he must continually leave behind, he cannot take it with him. What man should leave for the earth he takes from it; what man should take with him, what he has made useless for it, this man gives over to the earth with his earthly dust, thus causing the earth immense suffering in its entire life, in its whole collective being.

It is indeed the case that at first, just through spiritual vision, something weighs heavily upon the human soul, something like a tremendous feeling of tragedy. And only when one surveys wider epochs of time, when one beholds the development of entire systems, only then is the prospect revealed that, when the earth will have approached its end, in later stages of human evolution — in the Jupiter, Venus, Vulcan stages — will man be able to restore the balance, to annul the debt.

Thus it is not only by passing through the experiences of a single life that man fashions karma, but man creates karma, world karma, cosmic karma, just through the fact that he is an earthly human being, that he is an inhabitant of the earth, and draws his substance from the earth.

Here it is possible to look away from man, to look towards the rest of nature and see how — though man must burden himself with the debt of which I have just told you — balance is nevertheless continually restored by cosmic beings. And here one penetrates into wonderful secrets of existence, into secrets which, when taken in conjunction with each other, become something from which one can first gain a conception of the wisdom of the world.

Let us turn our gaze away from man and towards something which has claimed much of our attention during the last few days, let us turn our gaze to the world of the birds, represented for us by the eagle. We spoke of the eagle as the representative of the bird-world, as the creature which synthesizes the characteristics and forces of the bird-kingdom. When we consider the eagle, we are in fact considering, in their cosmic connection, all the attributes which prevail in the bird-world as a whole. In future, therefore, I shall simply speak of 'the eagle'

I have told you how the eagle actually corresponds to the head of man, and how those *Cosmic* forces which give rise to thoughts in the human head give rise in the eagle to his plumage. So that the sun-irradiated forces of the air, the light-imbued forces of the air, are actually working in the eagle's plumage.

This is what shimmers in the eagle's plumage — the light-irradiated power of the air.

Now the eagle — to whom many bad qualities may certainly be ascribed — does nevertheless possess, as regards his cosmic being, the remarkable attribute that outside his skin, in the structure of his plumage, everything is retained which is formed in it by the sun-irradiated forces of the air. What takes place here is, in fact, only to be noticed when the eagle dies.

For it is only when the eagle dies that one becomes aware of what a remarkable superficial digestion he has compared with the thorough-going digestion of the cow, with its process of chewing the cud. The cow is really the animal of digestion — again as representative of many creatures of the animal kingdom. Here digestion is thoroughly performed. The eagle, like all birds, digests in a superficial way; the business of digestion is only begun. In the eagle, compared with his whole existence, digestion is merely a subsidiary process and is treated as such. On the other hand, everything in the eagle which has to do with plumage proceeds in a thorough way. (In the case of some other birds this is even more so.) Everything to do with the feathers is worked out with immense care. Such a feather is indeed a wonderful structure. Here we find most strongly in evidence what may be called earthly *substance,* which the eagle has taken from the earth, *Cosmicized* by the forces of the heights, but in such a way that the eagle does not assimilate it; for the eagle makes no claim to reincarnation. He need not, therefore, be troubled about what is being brought about in the earthly *substance* of his plumage through the *Cosmic* forces of the heights; he need not be troubled about how this works on in the spiritual world.

Now, when the eagle dies and his feathers fall into decay — as already mentioned this holds good for every bird — the *Cosmicized* earthly substance ascends into spirit-land and becomes changed back into *Cosmic* substance.

You see we have a remarkable relative interplay as regards the relationship of our head to the eagle. What we cannot do, the eagle can; he can continually conjure forth from the earth what becomes *Cosmicized* in the earth through *Cosmic* forces working on earthly substance.

This, too, is why we experience such a remarkable sensation when we observe an eagle in its flight. We feel him as something foreign to the earth, something which has more to do with the heavens than with the earth, although he draws his substance from the earth. But how does he do this? He obtains his substance in such a way that, as regards the earth, he is just a robber. For according to what may be called the ordinary, commonplace law of earth-existence no provision was made for the eagle to get anything. He becomes a robber; he steals his substance, as is done in all sorts of ways by the bird-kingdom as a whole. But the eagle restores the balance. He steals his material substance, but allows it to be *Cosmicized* by the forces which exist as Cosmic forces in the upper regions; and after death he carries off into spirit-land those *Cosmicized* earth -forces which he has stolen. With the eagles the *Cosmicized* earth-*substance* withdraws into spirit-land.

Now the life of animals also does not come to an end when they die. They have their significance in the universe. And the eagle in flight is only a symbol of his real being. He flies as Earthly eagle — Oh, but he flies further after his death! The *Cosmicized* Earthly *substance* of the eagle nature flies into the universe in order to unite itself with the *Cosmic* substance of spirit-land.

You see what wonderful secrets of the universe one comes upon when one enters into the reality of these things. Only then does one really learn why the various animal and other forms of the earth are there. They all have their great, their immense significance in the whole universe.

And now let us turn to the other extreme, to something which we have also studied during these days, let us turn to the cow, so venerated by the Hindu. There we have the opposite extreme. Just as the eagle is very similar to the head, so is the cow very similar to the human digestive system. The cow is the animal of digestion. And, strange as it sounds, this animal of digestion consists essentially of *Cosmic* substance into which the *Earthly Forces that are* consumed is merely scattered and diffused. In the cow is the *Cosmic* substance and everywhere the *Earthly Forces* penetrates into it, and is absorbed, made use of by the *Cosmic* substance. It is in order that this may happen in a really thorough way that the process of digestion in the cow is so comprehensive, so fundamental. It is really the most fundamental digestive process that can be conceived, and in this respect — if I may put it so — the cow fosters what is fundamental to animal nature more thoroughly than any other animal in the absolute sense. She actually brings animal-nature — this animal egoism, this animal egoity — out of the universe down on to the earth, down into the region of earth-gravity.

No other animal has the same proportion between the blood-weight and the entire body-weight as the cow; other animals have either less or more blood than the cow in proportion to the weight of the body. And weight has to do with gravity and the blood with egoity; not with the ego, for this is only possessed by man, but with egoity, with separate existence. The blood also makes the animal, animal — the higher animal at least. And I must say that the cow has solved the world-problem as to the right proportion between the weight of the blood and the weight of the whole body — when there is the wish to be as thoroughly animal as possible.

You see, it was not for nothing that the ancients called the zodiac 'the animal circle'. The zodiac is twelvefold; it divides its totality into twelve separate parts. Those forces, which come out of the cosmos, from the zodiac, take on form and shape in the animals. But the other animals do not conform to the zodiacal proportion so exactly. The cow has a twelfth part of her body-weight in the weight of her blood. With the cow the blood-weight is a twelfth part of the body-weight; with the donkey only the twenty-third part; with the dog the tenth part. All the other animals have a different proportion. In the case of man the blood is a thirteenth of the body-weight.

You see, the cow has seen to it that, in her weight, she is the expression of animal nature as such, that she is as thoroughly as possible the expression of what is cosmic. A fact I have mentioned repeatedly during these days — namely that one sees from the astral body of the cow that she actually manifests something lofty in *its* material substance. This comes to expression of itself through the fact that the cow maintains the partition into twelve in her own inner relationships of weight. The cosmic in her is at work. Everything to do with the cow is of such a nature that the forces of the earth are working into *Cosmic* substance. In the cow earth-heaviness is obliged to distribute itself according to zodiacal proportion. Earth-heaviness must accommodate itself to allow a twelfth part of itself to fall away into egoity. What the cow possesses as *Cosmic* substance has necessarily to enter into earthly conditions.

Thus the cow, lying in the meadow, is in actual fact *Cosmic* substance, which earth-*substance* takes up, absorbs, makes similar to itself.

When the cow dies, this *Cosmic* substance which the cow bears within herself can be taken up by the earth, together with the earthly *forces*, for the well-being of the life of the whole earth. And man is right when he feels in regard to the cow: You are the true beast of sacrifice, for you continually give to the earth what it needs, without which it could not continue to exist, without which it would harden and dry up. You continually give *Cosmic* substance to the earth, and renew the inner mobility, the inner living activity of the earth.

When you behold on the one hand the meadow with its cattle, and on the other hand the eagle in flight, then you have their remarkable contrast: the eagle who, when he dies, carries away into the expanses of spirit-land that earth *substance,* which — because it is *Cosmicized* — has become useless for the earth; and the cow, who, when she dies, gives to the earth *cosmic* matter and thus renews the earth. The eagle takes from the earth what it can no longer use, what must return into spirit-land. The cow carries into the earth what the earth continually needs as renewing forces from spirit-land.

Here you become aware of something like an upsurging of feelings and perceptions from out of initiation-science. It is usually believed about this initiation-science, well, that one certainly studies it, but that it results in nothing but concepts, ideas. One fills one's head with ideas about the super-sensible, just as one otherwise fills one's head with ideas about the things of the senses. But this is not how it is. Penetrating ever further into this initiation-science, we reach the point of drawing forth from the depths of the soul feelings and perceptions, the existence of which we formerly did not even surmise, but which nevertheless are there unconsciously in every human being; we reach the point of experiencing all existence differently from the way we experienced it before. And so I can describe to you an experience which actually belongs to the living comprehension of spiritual science, of initiation science. It is an experience which would make us acknowledge that if man alone were upon the earth, we should — if we recognize his true nature — have to despair of the earth ever receiving what it needs, namely, that at the right time *Cosmicized Earthly Substance* should be withdrawn and spirit-substance bestowed. We should have to experience an opposition between man and the being of the earth, which causes great, great pain, and causes that pain because we have to admit that, if man is to be rightly man upon the earth, the earth cannot be rightly earth because of man. Man and earth have need of each other, but man and earth cannot mutually support each other. What the being of the one requires is lost to the other; what the other needs is lost to the one. And we should have no security as regards the life-relationship between man and earth, were it not that the surrounding world enables us to say: What the human being is unable to achieve as regards the carrying of *Cosmicized* earth-substance over into spirit-land, this is accomplished by the bird-kingdom; and what man is unable to do as regards giving *Cosmic* substance to the earth, this is accomplished by the animals which chew the cud, as represented by the cow.

In this way, you see, the world is rounded into a whole. If we look only at man, uncertainty enters our feelings as regards the being of the earth; if we look at what surrounds man our feeling of certainty is restored.

And now you will wonder even less that a religious world-conception, which penetrates so deeply into the spiritual as does Hinduism, venerates the cow, for she is the animal which continually spiritualises the earth, which continually gives to the earth that *Cosmic* substance which she herself takes from the cosmos. And we must learn to accept as actual reality the picture that, beneath a grazing herd of cattle, the earth below is quickened to joyful, vigorous life, that there below the elemental spirits rejoice, because they are assured of their nourishment from the cosmos through the existence of the creatures grazing above them. And we would have to make another picture of the dancing, rejoicing airy circle of the elemental spirits hovering around the eagle. Then again one would portray spiritual realities, and in the *Cosmic* realities one would see the *Earthly*; one would see the eagle extended outwards in his aura, and playing into the aura the rejoicing of the elemental air-spirits and fire-spirits of the air.

And one would see that remarkable aura of the cow, which so strongly contradicts her earthly nature, because it is entirely cosmic; and one could see the lively merriment in the senses of the elemental earth-spirits, who are thus able to perceive what has been lost to them because they are sentenced to live out their existence in the darkness of the earth. For these spirits what here appears in the cows is sun. The elemental spirits, whose dwelling place is in the earth, cannot rejoice in the Earthly sun, but they can rejoice in the astral bodies of the animals which chew the cud.

For, indeed, if one looks upon the world as these one cannot avoid the conclusion that eagles in their flight have no purpose, apart from the fact that they can be used in making armorial crests; cows are Earthly useful because they give milk, and so on. But because man also is regarded only as a Earthly being, he only possesses Earthly usefulness; and all this has no meaning for the world as a whole.

If people are unwilling to go further than this, they will certainly not reach the level where a world-meaning can appear; we must pass on to what the *Cosmic,* to what initiation-science can say to us about the world; then we shall certainly discover the meaning of the world. Then we shall find this meaning of the world as we discover wonderful mysteries in all existence — mysteries such as that which unfolds itself in connection with the dying eagle and the dying cow; and there between them the dying lion, which in his turn so holds *Cosmic* substance and *Earthly* substance in balance within himself, through the harmony he establishes in the rhythm of breathing and of blood, that it is he who regulates, through his group-soul, how many eagles are necessary, and how many cows are necessary, to enable the correct process both upwards and downwards to take its course in the way I have described to you.

You see, the three animals, eagle, lion, ox or cow, they were created out of a wonderful intuitive knowledge. Their connection with man is imbued with feeling. For the human being, when he sees into the truth of these things, must really admit: The eagle takes from me the tasks which I myself cannot fulfil through my head; the cow takes from me the tasks which I myself cannot fulfil through my metabolism, through my limb system; the lion takes from me those tasks which I myself cannot fulfil through my rhythmic system. And thus from myself and the three animals something complete is established in the cosmos.

Thus one lives one's way into cosmic relationships. Thus one feels the deep connections in the world, and learns to know how wise are those powers which hold sway in the world of being into which man is woven, and which live and move around him.

In this way, you see how we were able to weld together into a whole the diverse matters which came to our knowledge when we sought to discover man's connection with the three animal representatives about whom we have spoken in recent weeks.   (9)

These lectures deal with the inner connection between appearance and reality in the world, and you have already seen that there are many things of which those whose vision is limited to the world of appearance have no idea. We have seen how every species of being — this was shown by a number of examples — has its task in the whole nexus of cosmic existence. Now today, as a kind of recapitulation, we will again consider what I said recently about the nature of several beings and in the first place of the butterfly. In my description of this butterfly nature, as contrasted with that of the plants, we found that the butterfly is essentially a being belonging to the light — to the light in so far as it is modified by the forces of the outer planets, of Mars, of Jupiter, and of Saturn. Hence, if we wish to understand the butterfly in its true nature, we must in fact look up into the higher regions of the cosmos, and must say to ourselves: These higher cosmic regions endow and bless the earth, with the world of the butterflies.

The bestowal of this blessing upon the earth has an even deeper significance. Let us recall how we had to say that the butterfly does not participate in what is directly connected with earthly existence, but only indirectly, in so far as the sun, with its power of warmth and light, is active in this earthly existence. Actually a butterfly lays its eggs only where they do not become separated from sun activity, so that the butterfly does not entrust its egg to the earth, but only to the sun. Then out creeps the caterpillar, which is under the influence of Mars-activity, though naturally the sun influence always remains present. Then the chrysalis is formed, and this is under the influence of Jupiter-activity. Out of the chrysalis emerges the butterfly, which can now in its iridescent colours reproduce in the earth's environment the luminous Sun-power of the earth united with the power of Saturn.

Thus in the manifold colours of the butterfly world we see, in the environment of earth-existence, the direct working of Saturn-activity within the sphere of the earthly. But let us bear in mind that the substances necessary for earth-existence are in fact of two kinds. We have the purely material substances of the earth, and we have the *Cosmic* substances; and I told you that the remarkable thing about this is that in the case of man the underlying substance of his metabolic and limb system is *Cosmic* whereas that of the head is *Earthly.* Moreover in man's lower nature *Cosmic* substance is permeated with the activity of *Earthly* forces, with the action of gravity, with the action of the other earthly forces. In the head, the earthly substance, conjured up into it by the whole digestive process, the circulation, nerve-activity and the like, is permeated by super-sensible *Cosmic* forces, which are reflected in our thinking, in our power of forming mental pictures. Thus in the human head we have *Cosmicized Earthly substance,* and in the metabolic-limb-system we have earthized — if I may coin a word — earthized *Cosmic* substantiality.

Now it is this *Cosmic substance* that we find to the greatest degree in the butterfly. Because a butterfly always remains in the sphere of sun-existence, it only takes to itself earthly matter — naturally I am still speaking pictorially — as though in the form of the finest dust. It also derives its nourishment from those earthly substances which are worked upon by the sun. It unites with its own being only what is sun-imbued; and it takes from earthly substance only what is finest, and works on it until it is entirely *Cosmicized*. When we look at a butterfly's wing we actually have before us earthly *substance* in its most *Cosmicized* form. Through the fact that the matter of the butterfly's wing is imbued with colour, it is the most *Cosmic* of all earthly substances.

The butterfly is the creature which lives entirely in *Cosmicized* earth-matter. And one can even see spiritually how in a certain way a butterfly despises the body which it carries between its coloured wings, because its whole attention, its whole group-soul being, is centred on its joyous delight in the colours of its wings.

And just as we marvel at its shimmering colours as we follow it, so also can we marvel at its own fluttering joy in these colours. This is something which it is of fundamental importance to cultivate in children, this joy in the spirituality fluttering about in the air, which is in fact fluttering joy, joy in the play of colours. The nuances of butterfly-nature reflect all this in a wonderful way: and something else lies in the background as well.

We were able to say of the bird — which we regarded as represented by the eagle — that at its death it can carry *Cosmicized* earth-substance into the spiritual world, and that thereby, as bird, it has the task in cosmic existence of *Cosmicizing* earthly matter, thus being able to accomplish what cannot be done by man. The human being also possesses in his head earth-matter which has been to a certain degree *Cosmicized,* but he cannot take this earthly matter into the world in which he lives between death and a new birth, for he would continually have to endure unspeakable, unbearable, devastating pain, if he were to carry this *Cosmicized* earth-matter of his head into the Cosmic world.

The bird-world, represented by the eagle, can do this, so that thereby a connection is actually created between what is earthly and what is extra-earthly. Earthly matter is, as it were, gradually converted into spirit, and the bird-creation has the task of giving over this *Cosmicized* earthly matter to the universe. One can actually say that, when the earth has reached the end of its existence, this earth-matter will have been *Cosmicized,* and that the bird-creation had its place in the whole economy of earthly existence for the purpose of carrying back this *Cosmicized* earth-matter into spirit-land.

It is somewhat different with butterflies. The butterfly *Cosmicizes* earthly *substance* to an even greater degree than the bird. The bird after all comes into much closer contact with the earth than does the butterfly. I will explain this in detail later. Because the butterfly never actually leaves the region of the sun, it is in a position to *Cosmicize* its *substance* to such a degree that it does not, like the bird, have to await its death, but already during its life it is continually restoring *Cosmic substance* to the environment of the earth, to the cosmic environment of the earth.

Only think of the magnificence of all this in the whole cosmic economy! Only picture the earth with the world of the butterflies fluttering around it in its infinite variety, continually sending out into world-space the *Cosmicized* earthly *substance* which this butterfly-world yields up to the cosmos! Then, with such knowledge, we can contemplate the region of the world, of the butterflies encircling the earth with totally different feelings.

We can look into this fluttering world and say: From you, O fluttering creatures, there streams out something still better than sunlight; you radiate spirit-light into the cosmos! Our materialistic science pays but little heed to things of the spirit. And so this materialistic science is absolutely unequipped with any means of grasping at these things, which are, nevertheless, part of the whole cosmic economy. They are there, just as the effects of Earthly activities are there, and they are even more real. For what thus streams out into spirit-land will work on further when the earth has long passed away, whereas what is taught by the modern chemist and physicist will reach its end with the conclusion of the earth's existence. So that if some observer or other were to sit outside in the cosmos, with a long period of time for observation, he would see something like a continual outstreaming into spirit-land of *substance* which has become *Cosmicized,* as the earth radiates its own being out into cosmic space; and he would see — like scintillating sparks, sparks which ever and again flash up into light — what the bird-kingdom, what every bird after its death sends forth as glittering light, streaming out into the universe in the form of rays: a shimmering of the spirit-light of the butterflies, and a sparkling of the spirit-light of the birds. (10)

Another step behind the Veil

You may have noticed on page 116 "so that he doesn't permeate the bodily life in the right way with the *Cosmic Forces* **active in the head — the astral, the ego-being** ". *I put this in bold to highlight another layer to RS story. Somewhere RS comments that we have all four bodies working in every circumstance. In the story we have had in front of us, we have in the nerve sense system, the Spirit inspired Cosmic Forces working with the Earthly inspired Earthly Substance, while in the metabolism we have the Astrally inspired Cosmic Substance working with the Etheric inspired Earthly Forces. We have had 2 players in each system.*

RS's comment regarding all 4 activities in every circumstance means, we need to find the four activities, within each of the 3 Earthly organisations. So within the nerve sense activity we have all four activities, not two. So when he says Spirit inspired Cosmic Forces he means the Spirit is the dominate partner, and the Astral activity is present but secondary. Similarly the Earthly inspired Earthly Substance, actually means the Earthly is dominate and the Etheric is a secondary process.

Within the metabolic system the Astrally inspired Cosmic Substance, means the Astral body is the dominate actor providing movement to the metabolic processes. However the Spirit is present providing direction. The Etheric inspired Earthly Forces, means the Etheric is dominate while the Earthly is offering support.

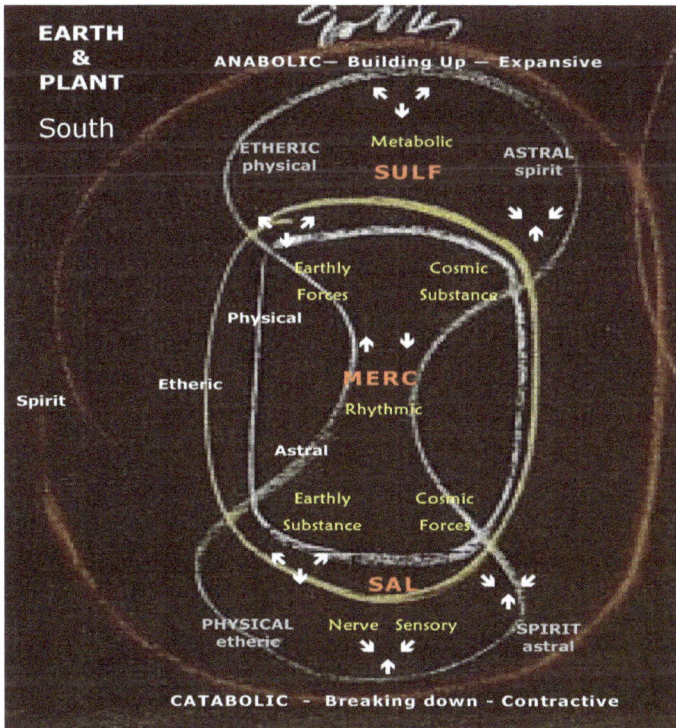

185

Four New Preparations

Sand

River silica sand, not beach calcium sand

Add to a clay soil at 2 – 3 cms deep, work it in. Add as needed.

Homeo – Triturate in Lactose for an hour, then water potencies

In horn mid summer to mid winter

Brings in the Cosmic Substance and holds the Cosmic Forces in the ground

*Activates the Internal **Astral** and Spirit in the Metabolic system*

Clay

In a horn from mid winter to mid Summer to emphasis it rising capacity

Could also do 6 wks either side of mid winter

Apply to sandy soil. A 10 cm ball per 10 cubic meters of garden

Homeo – Triturate in Lactose for an hour then water potencies

Stimulates the upward movement of the Cosmic Forces

*Activates the Internal **Spirit** and Astral in the Nerve system*

Humus

Applied as compost .

Homeo Triturate in Lactose for an hour then water potencies

In horn on the Etheric cycle above the horizon Spring to Autumn

Stimulates the Etheric activity of the Earthly Forces

*Activates the Internal **Etheric** and Physical in the Metabolic system*

Lime - *alkalis*

Ph of the soil – apply alkalis

Homeo Calcium Carbonate other alkais mixes? Magnesium (dolomite) all the cell salts. Triturate in Lactose for an hour then water potencies

In horn through Autumn till Spring

Draws the Earthly Forces and Etheric into the Earth

*Activates the Internal **Physical** and Etheric in the Nerve system*

References

1) Agriculture Course   GA 327

2) from the 1920 medical lectures

3) Human Questions Cosmic Answers 2 July 1923 GA 213

4) Driving Forces in World History 22 March 1923 GA 222

5) The Groups souls of Animals, Plants and Minerals, 2 February 1908

6) Cosmic Workings in Earth and Man, lecture 5 , 31 October 1923

7) Physiology and Therapeutics, Lecture 3, 9 October 1920

8) The Relation of Man to the 3 Worlds 23 august 23 GA 227

9) Man as Symphony Lecture 3 , 21 October 1923, Dornach

10) Cosmic Workings in Earth and Man, 27 October 1923, Dornach

Diagrams

Glen Atkinson

Rudolf Steiner

Web addresses

Glenopathy   www.garudabd.org

BdMax   www.bdmax.co.nz

Mark Moodie Publications   https://moodie.biz

When is 'Seed Chaos'

RS is in Italics

Introduction

Within Dr Steiner's Biodynamics there is a quirky 'idea' entitled 'Seed Chaos'. This is the period in a plants growth when the Star forces, which carry the 'architectural' species impulse of a plant, joins with the growing plant to direct the new seed and its subsequent plant, through its growth cycle. It is fair to ask, why is this important? It is important because if it is possible to influence the architectural plans of a species, as it enters, then it would be possible to make dramatic steps in plant development, such as having a perennial coriander plant, that will not rush off to seed immediately, or some other significant trait we might like. Biodynamic plant breeding has been taking place for many years. However there are only a few examples of plant development, mostly 'golden wheat' which has achieved such things as greater spacing between the seeds in the head. However there are no examples of dramatic changes such as annual plants becoming perennials and vice versa. So comparatively small changes have been achieved over long periods of time, which some might say are comparable to changes achieved through hybridisation.

It is now common to hear that this 'seed chaos' event takes place at germination. I understand Alex Podolinski promotes this, while Enzo Nastati, makes it the central premise of his '9 Introductory lectures in Biodynamic Agriculture'. Most of the 'big ideas' presented there, hang off this belief, and a series of chromatography pictures Alex produced. This collective belief, gives the 'moon planting' movement a 'glittering' philosophic justification, for the many scientific trials that show the time of sowing seed, has some influence upon the subsequent growth of the plant. A keen observer of plant growth, however will have noted that there are many such possible moments in a plants development. Whether there is light or not, whether it is a wet season or not, and what BD sprays are applied and when, all can be seen to have a significant effect on the subsequent plant development. So yes, sowing time does influence plant growth, but **is this the event that can influence the way in which a species expresses itself.** Is RS's description of this seed chaos event, talking about germination?

My study of RS comments regarding **'seed chaos',** shows he saw it as **occurring at fertilisation**. The ' money shot' is in the Agriculture Course, in the Discussion after Lecture IV — *"If there is fertilised seed at all, the chaos is complete."* (1) I would have thought this would be the end of the matter, however as we can see, the 'germination' belief is very much alive and well. Sadly this is not unusual in this movement. My website (2) has articles addressing various unsubstantiated 'BD beliefs' that are being passed from one generation to another, based upon 'the guru says'.

I have been concerned with 'other things, over recent years. So it is only now I have time to address this somewhat 'small' but vital question. During 2018, while addressing Enzo's lectures (3) and RS's Plant Growth story (4) I had a discussion with Stewart Lundy on this topic. He was very helpful with various lecture quotes and has compiled his version of that conversation into a essay entitled " Bringing Order to Seed Chaos". (5) He gives a more scholarly presentation than I, and provides a broader context of other instances of 'Diffused Chaos' discussed by RS. This is a good contribution to the conversation and well worth reading.

Texts

As the accompanying references show, I understand RS to be saying **Seed Chaos - when the immediate structural plans from the Stars, for the next cycle, are given to the plant, and inscribed as Spirit forces into its energetic matrix - occurs during the Pollination period of a plants growth cycle**. There are several quite direct statements in this regard.

The context of this discussion has to be placed within RS general understanding of how plants grow. My 'RS Plant Growth Story' (4) provides an overview. I identified 8 different parts of the story, all of which are based upon the fundamental observation that a plant can not just be seen as growing within the season that we see it in front of us. Much of what is in front of us this season, is the result of all that occurred in the previous growing season, along with what occurred during the Autumn and Winter just passed. The seed has quite a journey. Once it is pollinated, it goes through the ripening process, where some seeds develop defined primary leaves and root shoots, within themselves. They dehydrate and become stable while moving through the winter. RS calls this phase , when the seed is lingering in the Earth, the 'Fructification' time, before the upward Spring surge, sees the seed germinate and push upwards. The Light processes from above stimulate the leaf development up to the flowering, where DNA splitting and cell division / Seed Chaos occurs, and on the cycle goes. This story is told in several different ways in various lectures and I have attempted to summarise all the relevant lectures.

For the 'Seed Chaos' phase of this process we have the following from Lecture 2 of the Ag Course.

*"Now with regard to the cultivation of the soil there is a point of great importance which must be thoroughly understood. It is a point I have often dealt with amongst Anthroposophists. It is that we know the conditions which the forces of the cosmic spaces can work upon the earthly realm. Let us begin with **seed formation**. The **seed which gives rise to the embryo** of the plant is generally regarded as a molecular structure of exceptional complexity, and science lays great stress upon this interpretation. The molecules it is said have a certain structure, in simple molecules it is simple, in complicated molecules it becomes more and more complex, until we come to the extreme complexity of the albuminous or protein molecule. People stand in wonder and astonishment at the enormous complexity of the structure supposed to exist in the seed. They do so because they reason as follows. The albumen (or protein) molecule, they say, must be enormously complex, and since its structure was determined by the embryonic conditions of the seed, the latter's microscopic or ultra-microscopic content must also have a structure of enormous complexity. Well it is complex indeed in the beginning. As the earthly albumen is formed, its molecular structure is driven to the utmost complexity; but this alone would never give rise to a new organism. For the organism arising from the seed does not proceed by a mere continuation in the off- spring of what was present in the parent plant or animal.*

*What happens is that when the **embryonic structure** has reached the highest stage of complexity in the earth domain **it falls to pieces and becomes a "little chaos".** It breaks up and dissolves, one might say, into "world-dust".* (**DNA splitting**) *And when this little chaos of world-dust is there, the whole surrounding cosmos begins to work upon it to stamp it with its own image and to build up in it a structure conditioned by the forces of the Universe, working in upon it from every side (see drawing no. 3). Thus the seed becomes an image of the Cosmos. Every time this happens, and seed formation is carried through to the point of chaos, the new organism is built up from the seed-chaos by the activity of the cosmos. The parent organism has only the tendency to bring the seed to such cosmic position that through its affinity with this cosmic position **the cosmic forces will act in the proper direction so that, eg a dandelion will give rise to another dandelion and not a berberis.***

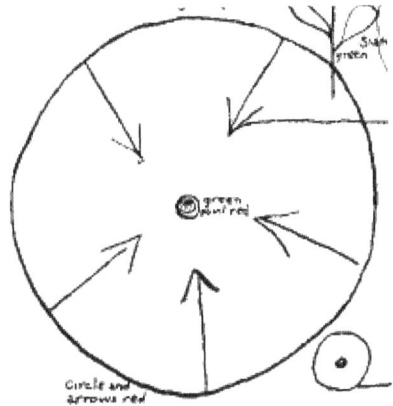

*But the new thing that is built up is always the image of some cosmic constellation. It is built up out of the cosmos. And if in the Earth we would make effective the forces of the cosmos, we must drive the earthly elements into the state of greatest possible chaos. This has to be the case whenever we want the cosmos to act upon our Earth. In the case of plant-growth this is in a certain sense provided for by nature herself. But just because every new organism is built up by the Cosmos it is necessary that the cosmic principles must be allowed freedom to work in the organisms **until the seed-formation is completed.***

*If for example, **we plant the seed of a given plant in the earth, the seed contains the impress of the whole cosmos from a particular cosmic direction , which means that it came under the influence of a particular constellation and received its particular form.** At the moment when the seed is placed in the soil it is strongly worked upon by **the terrestrial ("belly Ed.) forces,** and it is filled with the longing to deny the cosmic forces, in order that it may **spread and grow in all directions.** For the forces above the surface of the Earth do not want the plant to retain this cosmic form. The seed had to be driven to the point of chaos; but now that the plant is sprouting it is necessary to **oppose the terrestrial to the cosmic forces** which live as the form of the plant inside the seed. For the cosmic forces must be opposed and balanced, as it were, by the terrestrial forces. We must help the plant to become more akin to the Earth in its growth. This* *can only be done by introducing into the plant some form of living earthly matter* **which has not yet reached the state of chaos and seed formation**, *life which has been held up in a plant before the seeds have been formed. For this purpose a rich* **humus** *formation comes to man's assistance In those districts that are fortunate enough to possess it. Man can hardly find any artificial substitute for the fertility given to the soil by Nature through humus. What causes formation of humus? It arises from the absorption of remnants of living plants into the whole process of Nature, These remnants have not yet reached the state of chaos and respect the cosmic forces, as it were.* **If humus is used for the growth of plants the terrestrial forces are held fast within them.** (Earthly Forces) *The Cosmic Forces then work only* (with the help of clay) *in the upward stream, that terminates in seed-formation. While the terrestrial forces work in the development of flowers, leaf and so on, the cosmos only radiates its influence into all this..............* **Lec 2 Ag course — 1938 edition"**

Then in lecture 3 whilst talking of Hydrogen , the Spirits carrier, RS says *'and the other is when the hydrogen drives the basic element of the protein (for albumen)* **into the seed formation** *and there makes them independent of each other so that they become receptive of the influences of the Cosmos. In the tiny seed, there is chaos, and in the wide periphery of the Cosmos there is another chaos, and whenever the chaos at the periphery works upon the chaos within the seed, new life comes into being.*

A similar story is told in (6) however a bit abstractly. This is a story of the Warmth stream, that is sourced from the Stars and works inward through the Spheres. The other passages provided, talk of the specific Seed Chaos event, while here we have the Warmth stream physically manifesting within the plant fluids. It is this same Warmth Star stream, talked of as the functional activity in the Seed Chaos, but further down the story, as it moves through the plant, as Cambium and into the soil below. This whole stream is involved in the manifestation of the Star's plan throughout the plants life. The ripening process of the seed will be supported by the Cambium 'process'. Seed Chaos is a 'railway stop' along the way on the plants Star / World Spirit's journey, through Cambium and onto the centre of the Earth.

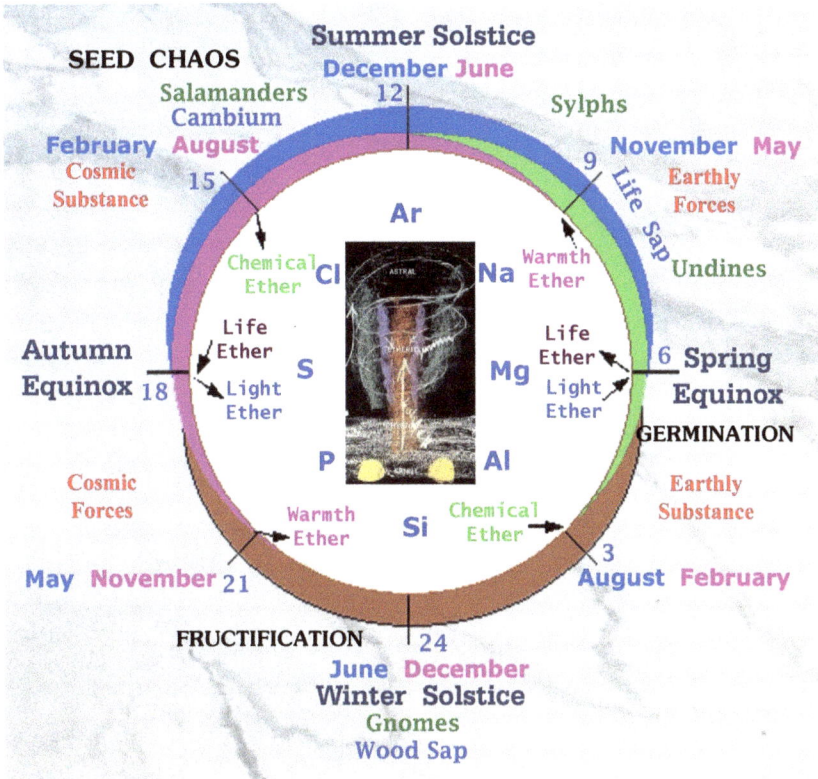

Summer Solstice

SEED   CHAOS   December June

Salamanders   12
Cambium

February August   Sylphs
Cosmic   15
Substance

Ar

Chemical Cl   Warmth Na Ether
Ether   Undines

Life Ether   Life Ether
S   Mg
Light Ether   Light Ether

Autumn Equinox   18   6 Spring Equinox

GERMINATION

P   Al

Cosmic Forces   Earthly Substance

Warmth Ether Si   Chemical Ether

May November 21   3 August February

FRUCTIFICATION   24

June December
Winter Solstice
Gnomes
Wood Sap

November May
Earthly Forces

Life Sap

This reference is a bit of a distraction, however I want to use this diagram, to provide you with the Big Picture, and the 'Sap Story' is an important part of that Story.

"When you look at the tree from above, you have first the pith inside: this gives the direction. Then layers of wood form round the pith. Towards the autumn the gum appears from the other side, and fastens the layers together. So we have the gummy wood of one year. In the next year this is repeated. Wood forms somewhere else, is again gummed together in the autumn, and so the yearly rings are formed. So you see everything clearly if only you understand that there are three things: wood sap, life sap, and cambium. The wood sap is the most fluid, it is really a chemical; the life sap is the giver of life; it is really, if I may so express myself, a living thing. And as for the cambium, there the whole plant is sketched out from the stars. It is really so. The wood sap rises and dies, then life again arises; and now comes the influence of the stars, so that from the thick, sticky cambium the new plant is sketched out. In the cambium one has a sketch, a sculptural activity. The stars model in it from the whole universe the complete plant form. So you see, we come from Life into the Spirit. What is modelled there is modelled from out of the World-Spirit. The earth first

193

gives up her life to the plant, the plant dies, the air environment along with its light once more gives it life, and the World Spirit implants the new plant form. This is preserved in the seed and grows again in the same way. "

*"You see, the wood sap is formed in the earthy-fluidic, the life sap in the fluidic-airy, and the cambium in the warm air, in the warm damp, or the airy-warmth. The plant develops warmth while it takes up life from outside. This warmth goes inward and develops the cambium inside. Or if the cambium does not yet develop, there is first of all developed a thicker substance: the plant gum. Plants form this plant gum in their inner warmth, and this, under certain conditions, is a powerful means of healing . This the plant gives back to the earth: Pitch, Resin, Amber. And if the plant retains it, it becomes cambium. Through the Wood sap the plant is connected with the earth; the life-sap brings the plant into connection with what circulates round the earth — with the airy-moist circumference of the earth. **But the cambium brings the plant into connection with the stars, with what is above, and in such a way that within this cambium the form of the next plant develops. This passes over to the seeds and in this way the next plant is born, so that the stars indirectly through the cambium create the next plant!** So that the plant is not merely created from the seed — that is to say, naturally it is created from the seed, but the seed must first be worked on by the cambium, that is: by the whole heavens."*

This Star process is taking place in the existing plant, coming from Above as a result of the plant growing well. Not when the seed germinates.

From another angle (7)

Something has always been taken for granted in this, namely that the molecules, as they are called, become more and more complicated the more we ascend from mineral, inorganic substance to organic substance. We say that the organic molecule, the cellular molecule, consists of carbon, oxygen, nitrogen, hydrogen, and sulfur. It is said that they are connected in some way but in a very complicated way. One of the ideals of natural science is to discover how these individual atoms in the complicated organic molecules are connected. Nevertheless, science admits that it will still be some time before we shall discover how one atom is connected with another within organic substance, within the living molecule. The mystery here, however, is this, that the more organic a substance is, the less one atom will be chemically connected with another, for the substances are whirled about chaotically, and even ordinary protein molecules, for instance in the nerve substance or blood substance, are in reality inwardly amorphous forms; they are not complicated molecules but inorganic matter inwardly torn asunder, inorganic matter that has rid itself of the crystallization forces, the forces that hold molecules together and connect the atoms with one another. This is already the case in the ordinary molecules of the organs, and it is most of all the case in the embryonic molecules, in the protein of the germ.

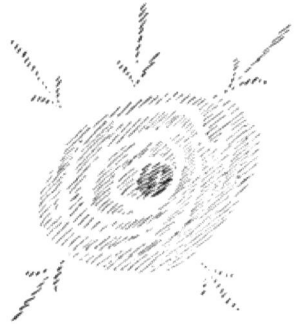

*If I draw the organism here (see drawing), and here the germ — and therefore the beginning of the embryo — the germ is the most chaotic of all as far as the conglomeration of material substance is concerned. This germ is something that has emancipated itself from all forces of crystallization, from all chemical forces of the mineral kingdom, and so on. Absolute chaos has arisen in this one spot, which is held together only by the rest of the organism. Because of the fact that here this chaotic protein has appeared, there is the possibility **for the forces of the entire universe to act upon this protein**, so that this protein is in fact **a copy of the forces of the entire universe**. Precisely those forces that then become formative forces for the etheric body and for the astral body are present in the female egg cell, without fertilization yet having taken place. **Through fertilization,** this formation also acquires the physical body and the I, the sheath of the I, and therefore that which constitutes the formation of the I. **This arises through fertilization,** and this here (see drawing) is a pure cosmic picture, is a picture from the cosmos, because the protein emancipates itself from all earthly forces and thus can be determined by what is extraterrestrial. In the female egg cell, earthly substance is in fact subject to cosmic forces. The cosmic forces create their own image in the female egg cell.*

Note the similarity of this story and the picture, to the story and picture told in the Ag course quote.

A similar story is told in a Lecture to young doctors in 1924 (8)

"Let us try to picture the plants. How do people proceed today when they picture the plants? There is the soil of the earth. The seed is pictured as being laid into the soil and then the plant grows out of this. People are naive enough to think as follows: Hydrogen is a very simple molecule, consisting of two atoms. All kinds of things are imagined to form combinations. Alcohol is certainly a very complicated molecule. Carbon is there combined with hydrogen and oxygen and then one has something more complex. And now there come still more complicated substances with more and more complicated molecules. There was a period during the eighties and nineties of the last century when the titles of these were very complicated, consisting of more

than three lines in length. Yes, the molecule has become terribly complicated! And now still more so. Then it becomes a seed, and a seed is a most highly complicated combination. Then the plant grows out of the seed. But all this is nonsense. The basis of the seed formation is, in reality, **that earthly matter tears itself away from the principle of structure and passes over into chaos, becomes chaotic, contains no more forces of matter in itself. Then, when no earthly structure is present, what is working out of the cosmos can assert itself.** *The cosmic declares its readiness to mirror the cosmic structure in the minute.* **In the seed formation the "nothingness" asserts itself over against the earthly and the cosmos works into the nothingness.**

Take a quartz crystal. It is an earthly thing. Why? Why is the quartz crystal an earthly thing, retaining its form really in a very pedantic, rigid way? The quartz gets its form from an inner force and if you break it apart with a hammer the single parts always retain the tendency to be six-sided prisms, self-contained, six-sided pyramids. This tendency is present. You can as little rid the quartz of this tendency as you can get pedantry out of a man who is pedantic by nature. You may atomize a pedantic person, but he will still remain pedantic. The quartz does not allow itself to come to the point where the cosmos can do anything with its forces. Therefore the quartz has no life. **If the quartz could be pulverized to such a degree that in the single fragments it no longer had the tendency to be governed, in the single fragment, by its own forces, something living and cosmic would grow out of the quartz.** *This is what happens in the formation of a seed. In the seed, matter is driven out to such a degree that the cosmos can intervene with its etheric forces. The world must be seen as a perpetual entering into chaos and again an emergence from chaos. What is contained in quartz also came at one time from the cosmos, but it remained at a standstill, has become Ahrimanic. It no longer exposes itself to the cosmic forces. As soon as anything enters into the realm of the living it must always pass through chaos.* (consider these comments in regard to Horn Silica 501)

This again is something which will help you to meditate in the sense of medicine. And you can also picture the developed plant – how it grows from leaf to leaf, and so on. You come to **the formation of the seed** *in the fruit. Whereas you otherwise picture the seed plant as brightness it now becomes dark, quite dark. Then again comes the light,* **when the forces from outside take hold.** *In this way, too, you can make an imaginative picture from the being of the plant. "*

Fructification

We can not talk of the Seed Chaos period without setting it into the whole plant cycle. Pollination can be seen as the 'male' reproductive phase, but is only the start of the new plants journey. It reaches a certain fulfillment when it combines with the 'female' Earthly growth forces active during the winter, and reaches its birth point at germination.

This story is best told in Lecture 7 of Man as a Symphony of the Creative World (9)

*"After it has passed through the sphere of the sylphs, the plant comes into the sphere of the elemental fire-spirits. These fire-spirits are the inhabitants of the warmth-light element. When the warmth of the earth is at its height, or is otherwise suitable, they gather the warmth together. Just as the sylphs gather up the light, so do the fire-spirits gather up the warmth and **carry it into the blossoms of the plants.***

*Undines carry the action of the chemical ether into the plants, sylphs the action of the light-ether into the plant's blossoms. And the pollen now provides what may be called little air-ships, to enable **the fire-spirits to carry the warmth into the seed.** Everywhere warmth is collected with the help of the stamens, and is carried by means of the pollen from the anthers to the seeds and the seed vessels. And **what is formed here in the seed-bud is entirely the male element which comes from the cosmos.** It is not a case of the seed-vessel being female and the anthers of the stamens being male. In no way does fructification occur in the blossom, but only the pre-forming of the male seed. **The fructifying force is what the fire-spirits in the blossom take from the warmth of the world-all as the cosmic male seed, which is united with the female element.** This element, drawn from the forming of the plant has, as I told you, already earlier seeped down into the ground as ideal form, and is resting there below. **For plants the earth is the mother, the heavens the father.** And all that takes place outside the domain of the earth is not the mother-womb for the plant. It is a colossal error to believe that the mother-principle of the plant is in the seed-bud. The fact is that this is the male-principle, which is drawn forth from the universe with the aid of the fire-spirits. The mother comes from the cambium, which spreads from the bark to the wood, and is carried down from above as ideal form. And what now results from **the combined working of gnome-activity and fire-spirit activity — this is fructification.** The gnomes are, in fact, the spiritual midwives of plant-reproduction. **Fructification takes place below in the earth during the winter,** when the seed comes into the earth and meets with the forms which the gnomes have received from the activities of the sylphs and undines*

and now carry to where these forms can meet with the fructifying seeds.

*You see, because people do not recognize what is spiritual, do not know how gnomes, undines, sylphs and fire-spirits — which were formerly called salamanders — weave and live together with plant-growth, there is complete lack of clarity about the process of fructification in the plant world. There, outside the earth nothing of fructification takes place, but **the earth is the mother of the plant-world, the heavens the father**. This is the case in a quite literal sense. Plant-fructification takes place through the fact that the gnomes take from the fire-spirits what the fire-spirits have carried into the **seed bud as concentrated cosmic warmth** on the little airships of the anther-pollen. Thus the fire-spirits are the bearers of warmth.*

*And now you will easily gain insight into the whole process of plant-growth. First, with the help of what comes from the fire-spirits, the **gnomes down below instill life into the plant and push it upwards**. They are the fosterers of life. They carry the life-ether to the root — the same life-ether in which they themselves live. The undines foster the chemical ether, the sylphs the light-ether, the fire-spirits the warmth ether. And then the fruit of the warmth-ether again unites with what is present below as life. Thus the plants can only be understood when they are considered in connection with all that is circling, weaving and living around them. And one only reaches the right interpretation of the most important process in the plant when one penetrates into these things in a spiritual way.*

*When once this has been understood, it is interesting to look again at that memorandum of Goethe's where, referring to another botanist, he is so terribly annoyed because people speak of the eternal marriage in the case of the plants above the earth. Goethe is affronted by the idea that marriages should be taking place over every meadow. This seemed to him something unnatural. In this Goethe had an instinctive but very true feeling. He could not as yet know the real facts of the matter, nevertheless he instinctively felt that fructification should not take place above in the blossom. Only he did not as yet know what goes on down below under the ground, he did not know that **the earth is the mother-womb of the plants**. But, that the process which takes place above in the blossom is not what all botanists hold it to be, this is something which Goethe instinctively felt.*

You are now aware of the inner connection between plant and earth. But there is something else which you must take into account.

You see, when up above the fire-spirits are circling around the plant and transmitting the anther-pollen, then they have only one feeling, which they have in an enhanced degree, compared to the feeling of the sylphs. The sylphs experience their self, their ego, when they see the birds flying about. The fire-spirits have this experience, but to an intensified degree, in regard to the butterfly-world, and indeed the insect-world as a whole. And it is these fire-spirits which take the utmost delight in following in the tracks of the insects' flight so that they may bring about the distribution of warmth for the seed buds. In order to carry the concentrated warmth, which must descend into the earth so that it may be united with the ideal form, in order to do this the fire-spirits feel themselves inwardly related to the butterfly-world, and to the insect-creation in general. Everywhere they follow in the tracks of the insects as they buzz from blossom to blossom. And so one really has the feeling, when following the flight of insects, that each of these insects as it buzzes from blossom to blossom, has a quite special aura which cannot be entirely explained from the insect itself. Particularly the luminous, wonderfully radiant, shimmering, aura of bees, as they buzz from blossom to blossom, is unusually difficult to explain. And why? It is because the bee is everywhere accompanied by a fire-spirit which feels so closely

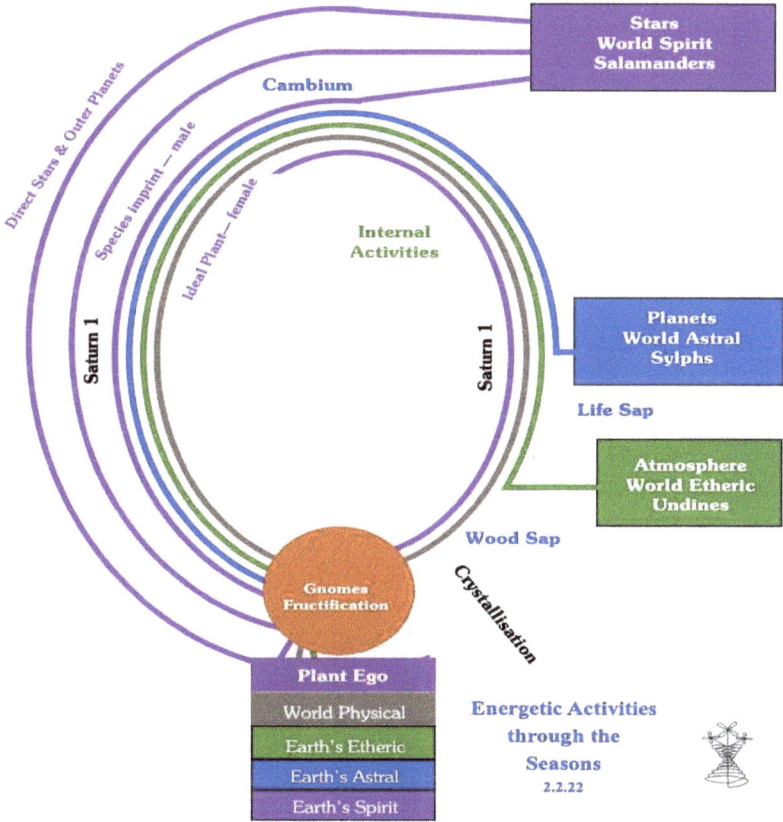

Stars
World Spirit
Salamanders

Cambium

Direct Stars & Outer Planets

Species Imprint – male

Ideal Plant – female

Internal
Activities

Saturn 1

Saturn 1

Planets
World Astral
Sylphs

Life Sap

Atmosphere
World Etheric
Undines

Wood Sap

Crystallisation

Gnomes
Fructification

Plant Ego

World Physical

Earth's Etheric

Earth's Astral

Earth's Spirit

Energetic Activities
through the
Seasons
2.2.22

related to it that, for spiritual vision, the bee is surrounded by an aura which is actually a fire-spirit. When a bee flies through the air from plant to plant, from tree to tree, it flies with an aura which is actually given to it by a fire-spirit. The fire-spirit does not only gain a feeling of its ego in the presence of the insect, but it wishes to be completely united with the insect.

Through this, however, insects also obtain that power about which I have spoken to you, and which shows itself in a shimmering forth of light into the cosmos. They obtain the power completely to spiritualize the physical matter which unites itself with them, and to allow the spiritualized physical substance to ray out into cosmic space. But just as with a flame it is the warmth in the first place which causes the light to shine, so, above the surface of the earth, when the insects shimmer forth into cosmic space what attracts the human being to descend again into physical incarnation, it is the fire spirits which inspire the insects to this activity, the fire-spirits which are circling and weaving around them. But if the fire-spirits are active in promoting the outstreaming of spiritualized matter into the cosmos, they are no less actively engaged in seeing to it that the concentrated fiery element, the concentrated warmth, goes into the interior of the earth, so that, with the help of the gnomes, the spirit-form, which sylphs and undines cause to seep down into the earth, may be awakened.

This, you see, is the spiritual process of plant-growth. And it is because the subconscious in man divines something of a special nature in the blossoming, sprouting plant that he experiences the being of the plant as full of mystery. The wonder is not spoiled, the magic is not brushed from the dust on the butterfly's wing. Rather is the instinctive delight in the plant raised to a higher level when not only the physical plant is seen, but also that wonderful working of the gnome-world below, with its immediate understanding and formative intelligence, the gnome-world which first pushes the plant upwards. Thus, just as human understanding is not subjected to gravity, just as the head is carried without our feeling its weight, so the gnomes with their light-imbued intellectuality overcome what is of the earth and push the plant upwards. Down below they prepare the life. But the life would die away were it not formed by chemical activity. This is brought to it by the undines. And this again must be imbued with light. And so we picture, from below upwards, in bluish, blackish shades the force of gravity, to which the impulse upwards is given by the gnomes; and weaving around the plant — indicated by the leaves — the undine-force blending and dispersing substances as the plant grows upwards. From above downwards, from the sylphs, light falls into the plants and shapes an idealized plastic form which descends, and is taken up by the mother-womb of the earth;

moreover this form is circled around by the fire-spirits which concentrate the cosmic warmth into the tiny seed-points. This warmth is also sent downwards to the gnomes, so that from out of fire and life, they can cause the plants to arise.

And further we now see that essentially the earth is indebted for its power of resistance and its density to the antipathy of the gnomes and undines towards amphibians and fishes. If the earth is dense, this density is due to the antipathy by means of which the gnomes and undines maintain their form. When light and warmth sink down on to the earth, this is first due to that power of sympathy, that sustaining power of sylph-love, which is carried through the air, and then to the sustaining sacrificial power of the fire-spirits, which causes them to incline downwards to what is below themselves. So we may say that, over the face of the earth, earth-density, earth-magnetism and earth-gravity, in their upward-striving aspect, unite with the downward-striving power of love and sacrifice. And in this inter-working of the downwards streaming force of love and sacrifice and the upwards streaming force of density, gravity and magnetism, in this inter-working, where the two streams meet, plant-life develops over the earth's surface. Plant-life is an outer expression of the inter-working of world-love and world-sacrifice with world-gravity and world-magnetism.

*From this you have seen with what we have to do when we direct our gaze to the plant-world, which so enchants, uplifts and inspires us. Here real insight can only be gained when our vision embraces the spiritual, the super-sensible, as well as what is accessible to the physical senses. This enables us to correct the capital error of materialistic botany, that fructification occurs above the earth. What occurs there is not the process of fructification, but **the preparation of the male heavenly seed for what is being made ready as the future Plant in the mother-womb of the earth.***

Germination is the plants birth. Conception is pollination, and 'pregnancy' is the journey through the autumn and winter to RS fructification, which has crystallisation, around mid winter, as ia pivotal point. Quoting Goethe, RS makes several comments that the 'fertility' of the plant, is not complete till after mid winter. Fructification is the culmination of the Gnomes task of combining the Cosmic Forces coming from above during last season, with the Earthly Substances, growth forces active within the Earth. The seeds are imbued with both these sets of forces, for the season ahead. This 'Fructified Seed', is then germinated / birthed in the soil. In most cases this is a simple, just add water process, for what is already there to be set in motion. RS talks then in lecture 2 of how the Earthly Forces above try and steal the Cosmic Forces away from their destiny…...and so the Cosmic and Earthly dance begins once more.

Conclusion

The real benefit of this clarification is what might be achieved in plant breeding. This requires further investigation, on my part. I commented that there are 7 different 'pathways' of activities RS describes. Each of these can be used as levers to bring the plant to a desired 'subjective' condition, that can influence the message the plant receives as the Star blue print. A separate essay is needed to outline the details of a way forward, for plant development.

"If there is fertilised seed at all, the chaos is complete."

References

(1)   https://wn.rsarchive.org/Lectures/GA327/English/BDA1958/Ag1958_discuss4.html

(2)   www.garudabd.org

(3)   http://garudabd.org/2019/03/27/in-response-to-enzo-nastati-intro-to-bd-in-9-meetings/

(4)   http://garudabd.org/dr-steiners-plant-story/

        https://docs.google.com/document/d/1hPurMuiCaBpkFKKo12ITiZHxW8VLbZmk7_nIKcnWqpU/edit

(6)   Cosmic Workings in Earth and Man Lecture 5 31 October 1923
        - http://wn.rsarchive.org/Lectures/Dates/19231031p01.html

(7)   Therapeutic Insights: Earthly and Cosmic Laws, Lecture 3 , July 1, 1921

(*)   Course for Young Doctors - Lecture 1 , Dornach, April 21, 1924

(9)   Man as Symphony of the Creative Word - Lecture 7, 2nd November, 1923
        http://wn.rsarchive.org/Lectures/Dates/19231102p01.

I trust your journey through these tales has been enjoyable.

For more - www.garudabd.org